T0296135

THE ORIGIN OF THE EARTH

THE ORIGIN OF THE EARTH

BY

W. M. SMART

M.A., D.Sc.

Regius Professor of Astronomy in the University of Glasgow

With Plates and Diagrams

CAMBRIDGE

AT THE UNIVERSITY PRESS

1953

CAMBRIDGE
UNIVERSITY PRESS

University Printing House, Cambridge CB2 8BS, United Kingdom

Cambridge University Press is part of the University of Cambridge.

It furthers the University's mission by disseminating knowledge in the pursuit of education, learning and research at the highest international levels of excellence.

www.cambridge.org
Information on this title: www.cambridge.org/9781107475403

First printed 1951
Second edition 1953
First published 1953
Re-issued 2014

A catalogue record for this publication is available from the British Library

ISBN 978-1-107-47540-3 Paperback

PREFACE

This book had its origin in a series of lectures delivered to members of the three fighting services during the last year of the Second World War and to various audiences on subsequent occasions. Preoccupations of many kinds have delayed the transmutation of the spoken word into the printed page but, undoubtedly, considerable advantage has accrued thereby since several topics of great importance, the fruits of post-war research, can now be included in the general discussion.

The book is intended to be, primarily, a record of scientific investigations in a subject of enthralling interest and we describe in the following pages the contributions of astronomy, physics, chemistry, geology and biology to the main theme inherent in the title of the book. That Science has not yet attained complete success in the field appropriate to its activities is not surprising; nevertheless, in the scientific story there are achievements which would have appeared incredible half a century ago.

Care has been taken to concentrate on the principal features of each particular topic and to resist the temptation to include details which, however important in themselves to the scientist, only serve to distract attention from the general argument.

The first nine chapters were completed in August 1948, but unavoidable delay in printing has allowed me to include in the final chapter brief references to new developments, some of a revolutionary character.

I am indebted to the Directors of the Lick Observatory, Mt Wilson Observatory, the Royal Greenwich Observatory and the Warner and Swasey Observatory for their kindness in allowing me to make use of some of their photographs, to the Oxford University Press for their permission to reproduce Plate IV from my book *Astronomy* and to the Staff of the Cambridge University Press for their care and attention in all matters of printing.

W. M. S.

UNIVERSITY OBSERVATORY
GLASGOW, W.2
June 1950

NOTE ON SECOND EDITION

In this edition a few residual errors and misprints have been corrected. The paragraph (p. 231) on *mesons* has been re-written.

W. M. S.

January 1953

CONTENTS

LIST OF PLATES

CHAPTER I

INTRODUCTION

To many it may seem presumptuous of anyone to attempt to unravel the mysteries surrounding the origin of the Earth and, in a wider sense, of the Universe itself; to others it may even seem irreverent that the great act of Creation should be regarded as a field for scientific speculation. In his pursuit of knowledge of the external world the scientist, however, is not held within precise limits of time, past or future; for example, the measure of progress in astronomy through the ages can be expressed, in some respects at least, in terms of the increasing capacity of man to see still further into the future and to peer still more deeply into the mists of the past. There may be—and I think that there is—a point in the remote past where an 'iron wall', of cosmic fabrication, shuts us out from scientific contemplation of the antecedent state and the primeval evolutionary history of the Universe; but, until this point is reached, scientists have a legitimate domain of exploration in which, indeed, many triumphs have already been won.

The doctrine of Evolution as it concerns the biological sciences is familiar, in its general aspects at least, to all intelligent people in this scientific age; it is not always realized that evolution is an active principle in the material Universe. The Sun itself, or any star, is slowly evolving, its physical and chemical characteristics undergoing, slowly but relentlessly, changes which find their eventual expression in the heat and light radiated from the Sun's surface into space. Occasionally there is a violent discontinuity in this orderly transformation; a 'new star' (or *nova*) appears suddenly in the firmament, rising from obscurity to become perhaps one of the most brilliant objects in the sky and, after a brief period of grandeur, returning to the insignificance of the pre-*nova* stage. In the phenomenon of evolution of the continuous kind the scientist can trace the progressive changes in development—and this applies also to the post-explosion changes in the life of a 'new star'—with perhaps as much certitude as is claimed in the biological sciences.

The probing of the past is a feature of many human activities: the detective, for example, is concerned with actions, words, motives referring to past weeks or months or even years; the historian deals with much longer temporal spans of human activity and is bounded in time only by the absence of such documentary evidence as is necessary for his full appreciation of the past; the archaeologist goes still further back in his investigations of early man and the contemporary state of civilization; the geologist and the biologist take an immense leap into antiquity which is surpassed only by the astronomer or cosmogonist. In comparison with the historian, archaeologist, geologist and biologist the astronomer's presumption is just a matter of degree. We are concerned, then, in this book with the study of the past, pressing as far back as our means permit, and perhaps coming in sight of a barrier, as already suggested, beyond which it would seem we cannot pass. Our survey will cover a wide range, from the almost unbelievably minute atomic constituents of matter and particles no bigger than a grain of sand to stellar systems of unimagined grandeur; however, our main preoccupation will be with the Earth, its fellow planets and the Sun.

To this task we apply the scientific method which consists in making innumerable observations of the external world, measuring all that our most accurate instruments can measure, reaching general conclusions on the basis of consistent and reliable evidence, formulating natural laws which testify to the reign of order in the Universe, testing these laws continuously by further observations and predictions, modifying, refining and simplifying as the occasion demands. The expression of the natural law is itself a process in evolution; what seems to be thoroughly established to-day is found to-morrow to be not the whole truth but perhaps only a significant part of it or perhaps only a close approximation. Innumerable examples in the history of science are readily called to mind of which the following suffice for the present. The law of 'conservation of mass' in chemical reactions was an article of scientific faith for many decades; for example, in the formation of carbon dioxide from carbon and oxygen by combustion, as in a fire, this law asserted that the total weight of carbon and oxygen taking part in the reaction was precisely equal to the weight of carbon dioxide produced; although it is true that the measurements made with the most delicate balance ever constructed are in full accordance with the law of conservation of mass, yet the reaction cannot be reduced to the simple terms stated, for the union of carbon with oxygen produces carbon dioxide *and heat*, the latter eluding the functional capacity of the balance. But in recent years mass has been identified with energy, and vice versa, and as heat is a form of energy the combination of carbon with oxygen is an example, not of the law of conservation of mass in the sense indicated previously, but of the great law of conservation of energy the expression of which in this particular instance is:

The total energy, before combustion, of the participating substances—that is, the energy-equivalent of the mass of carbon and oxygen entering into combination—is equal to the total energy produced—that is, the energy-equivalent of the mass of carbon dioxide resulting from combustion, together with the heat energy liberated.

Kepler's laws of planetary motion, with their subsequent generalization by Newton, provide another striking example. Kepler's first law states that a planet *P* (Fig. 1) moves in an elliptical path around the Sun *S*, the latter being situated at one focus of the ellipse. This law was deduced from a long series of observations of the planet Mars made mainly by the famous Danish astronomer Tycho Brahe. To-day we know that it applies to all the planets and this conclusion might well have been reached by Kepler if he had been in possession of the appropriate observational material such as he had in the case of Mars. The law itself may be said to be a generalization which can be submitted to the test of verification as regards planets for which Kepler had insufficient information and for such planets as Uranus, Neptune and Pluto discovered long after his time. Kepler's second law was concerned with the

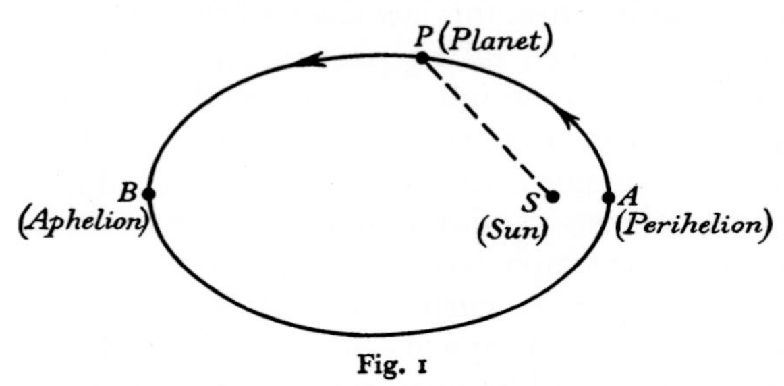

Fig. 1

rate at which the line *SP*, called the *radius vector*, joining the Sun to the planet rotated in the direction indicated by the arrow in Fig. 1; this rate is greatest when the planet is nearest the Sun at *A* (called *perihelion*) and least when the planet is furthest from the Sun at *B* (called *aphelion*).

The third law may be described most simply perhaps as follows. The Earth is of course a planet circulating around the Sun in an elliptic orbit such as is shown in Fig. 1 in the period we call a year; *AB* is called the *major axis* of the orbit and half this the semimajor axis; the length of the Earth's semimajor axis is used by astronomers as the astronomical unit of distance. If then T denotes the period (in years) required by any other planet, say Jupiter, to complete its path around the Sun, Kepler's third law in effect states that the semimajor axis of the planet's orbit (expressed in astronomical units) is equal to $T^{2/3}$. One of the significant facts about Kepler's three laws is that, for more than half a century, they were regarded as wholly independent of one another, with no conceivable bond uniting them in any way. The synthesis was effected by

Newton with the discovery of the famous law of gravitation which states that any particle of matter attracts any other particle with a force proportional to the product of the masses of the two particles and inversely as the square of the distance between them. From this single law the three laws of Kepler were easily deduced.

But the law of gravitation did more than introduce unification and simplification into the planetary field; it replaced observational generalizations by a physical principle—that of the power of a mass to attract any other mass—and it gave astronomers the means by which the position of any planet at any time could be accurately calculated and, its most sensational achievement, the means by which a planet hitherto unseen was added in the fullness of time to the solar family; to these achievements the laws of Kepler could in no real measure aspire. The discovery of the law of gravitation naturally evoked the question: By what means or process or mechanism can the apparent capacity of one chunk of the material universe for exerting the force of attraction on another chunk be explained? Through a further integration the theory of relativity, which asserts, it may seem mysteriously, that gravitation is simply a property of the four-dimensional world of space-time, allowed us to advance one step more in elucidating the mystery.

In the sequel we shall be concerned mainly with that aspect of the scientific outlook which relates to the inferring of antecedent conditions and processes of the material world from phenomena that have come under observation and from natural laws that have been established. The very remarkable achievements of scientists in the last few years have suggested to many lay minds that it is science alone that attains success through a close attention to the scientific method of description and interpretation, of seeking the causes that produce observed effects, of answering the questions 'how?' and 'why?' which are the expression of man's curiosity about the phenomena in his immediate or remote surroundings. It is well to realize, however, that the scientist is not unique in his approach to the satisfying of his intellectual curiosity; it is sufficient to mention the methods of the historian which have as much claim to the descriptive adjective 'scientific' as those of the atomic physicist occupied in producing that most lethal weapon, the uranium-bomb. Of course it is true that the historian is unable, as a rule, to advance step by step with the precision and confidence of the physicist; that, however, is not the fault of his method but of the paucity or incompleteness of the records on which he is forced to rely for his information. The historian, it is sometimes alleged, is prejudiced in favour of one summing-up of a particular phase of history rather than of another but, given honesty of purpose, such 'prejudice' is nothing more than the result of the weighing of motives and actions to the best of his ability according to the evidence available.

Let us not then exalt the scientific method unduly as the close preserve of the scientist nor, which is much more important, as the only means by which we attempt to discover the secrets of Nature. It is easy for the scientist to be a materialist if he sees only in the Universe the apparently relentless unfolding of natural law and forgets that there are domains where the laws of physics are irrelevant. But, more and more, scientists are realizing that they are exploring only one section of the great world of Nature in all its manifold complexity; beauty, moral conduct, spiritual values, religious experience are all outside his domain, yet all come necessarily within man's scrutiny when he attempts to interpret the Universe as a whole and strives to discern purpose therein. A great work of pictorial art could be analysed by the scientist in terms of chemical constitution, atomic and molecular structure, the laws of physical optics and all the rest; he might reduce Beethoven's Fifth Symphony to a collection of mathematical formulae in the theory of vibrations; in neither case would his interpretation be more than bare bones, incomplete and unsatisfying.

Even to the scientist the sense of beauty is as real as an atom or a star; he can explain the phenomenon of Saturn's Rings by means of complex formulae without ever having seen that wonder of the skies; but when he applies his eye to the telescope, he sees more than a collection of mathematical symbols, for then he sees beauty and, unless he be very dull indeed, he sees mystery and he experiences the feelings of awe and wonder. So too, for example, he can measure the relative positions of the two stars forming the twin-star Gamma Andromedae, he can ascertain their motions and calculate the masses and temperatures of the two stars and determine how far away the system is; but all these things do not specify the twin-star completely, for who can forget the beauty of colour—the rich yellow of one star and the wonderful blue of the other? Even in mathematics the appeal to the aesthetic sense is not as infrequent as may be supposed; I remember, just after the total solar eclipse of 1919 which did so much to support the new and strange theory of relativity and to arouse the intense interest of a war-weary world, discussing with the late Sir Arthur Eddington the *beauty* of Einstein's mathematical presentation; his face lit up with an enthusiasm and deep feeling such as many of us frequently experience in contemplating a moving work of art or in listening to heart-stirring music or in viewing a sublime spectacle of Nature.

It is suggested, then, that the reader should bear in mind the incompleteness of the picture which science gives of the beginnings of things; as we shall see, it can account in a surprisingly successful way for several elements in the story but it fails to discern any motive behind Creation, any omnipotent Mind, any guiding hand in the evolutionary process; that this is so is not a fault of the scientific method but of its limitations in a critical survey of the Universe from every possible angle. The story, then, we have to relate is only one aspect of the whole grand theme; perhaps we may go so far as to say that the conclusions of science in the present connexion, so far as these are definitive or even plausible, form the skeletal framework of the integrated organism, the flesh and living spark of which can only be perceived by eyes other than those of the scientist *qua* scientist.

Since man first acquired the power of rational thought he has speculated on the origin of the Universe as he knew it; in a sense, such speculations reflected the standard of his intellectual powers and of his knowledge of natural processes. At an early epoch of human history the association, however vague, of cause and effect with reference to the simplest of observed phenomena represented an important milestone in man's intellectual development; the planting of an acorn in the earth resulting in the growth of the oak, and the power of the Sun's heat to melt ice or snow are simple examples illustrating this relationship. But there was a class of phenomena, such as lightning and volcanic eruptions which could not be associated with an understandable natural origin and so the hypothesis of some supernatural agency was invoked to provide the link between cause and effect. It is not surprising then that as man, in successive ages, speculated on the origin of the Universe he was led to the acknowledgement of an all-creating Power who made 'the Heavens and the Earth and all that in them is'. The earliest known description of the Creation is the polytheistic account of the Babylonians about two thousand years before the beginning of the Christian era. Later came the superb account in the Book of Genesis with which most of us are familiar and which has dominated European theology and philosophy until comparatively recent times. As a result of the rapid march of science, especially in the last century, attention has inevitably been focused on the theme of the Biblical story and in the following pages we attempt to present a report, as it were, of what has been achieved.

Perhaps, here, we may ask legitimately if in probing, in the deepest sense, the mystery of Creation science has *really* been more successful than the poetic expounder of Hebrew cosmogony; the answer seems to be emphatically 'No'. Genesis begins with the majestic hypothesis of a divine Creator, implied in the words: 'In the beginning God....' As we shall see, the cosmogonist has on his part to postulate hypotheses which to him must be reasonable and conformable to established scientific laws. However far on the road of exploration these hypotheses take him, the ultimate goal seems to be as far out of sight as ever, although neighbouring landmarks continue to be investigated with ever-increasing thoroughness and understanding.

This book, then, deals mainly with the landmarks referred to. Although the whole grand synthesis has not been revealed to us by science—and possibly never will be—the story of the scientific adventure into the past is nevertheless an enthralling one and in its sum-total represents one of the loftiest achievements of the human mind.

From what has already been said in a general way about evolutionary processes, it can be accepted without cavil that the present state of the Earth, and of the Universe in general, represents a development from some antecedent state with marked differences as regards physical characteristics; for example, the fact that practically every cosmic body is dissipating heat (that is, energy) into space, with no substantial means of replenishment from external sources so far as we can see, is enough to suggest one aspect of this evolutionary process; in particular, the Earth is becoming cooler—very slowly, it is true, and at a rate that need not cause any serious perturbation to the human race for almost countless centuries—even allowing for the fact that by radioactive processes some of the energy locked up in the heart of the atom is being slowly released to counterbalance, but partially, the general tendency of our planet to lose heat by radiation into the voids of space.

In discussing the origin of the Earth we shall be concerned with three principal questions. The first question is: 'Whence came the Earth?'—or in the abbreviated form which we shall use: 'Whence?'. An alternative form, perhaps more suggestive of the problem set by the question, is: 'What is the Earth's parentage?' The question, in whatever form we use it, suggests that in the distant past the Earth was very different from what it is at present, that the material of which our planet is composed perhaps formed part of another cosmic body, and that somehow the Earth is a result—or, at any rate, one result—of some immense metamorphosis of part of the material Universe, perhaps explicable in terms of natural processes which are consonant with established scientific principles or perhaps explicable only, in the last resort, by the recognition of Creative Power. The problem of the origin of the Earth is but one aspect of a much wider problem, namely, the origin of the Sun's family of planets (Mercury, Venus, Earth, Mars, Jupiter, Saturn, Uranus, Neptune and Pluto), their attendant satellites (or moons), the vast horde of minor planets and other bodies associated with the Solar System. Nor is the problem eventually circumscribed even by these inquiries, for our survey will touch on the formation and development of stars—the Sun, in particular—and the evolution of the Universe itself. In Chapters II–V we shall attempt to answer the question 'Whence?', including in these chapters such preliminary information as is necessary for an understanding of the arguments.

The second question is 'When?', or in extended form, 'When was the Earth formed as a planetary entity?', or more briefly, 'How old is the Earth?' This stage of our inquiry will introduce us to the celebrated controversies of the late Victorian decades when the conclusions of geology and biology on the one hand appeared irreconcilable with the conclusions of physical science on the other, in discussing the problem of the Earth's age. We shall be further introduced to the fascinating developments in physics during the present century which were inaugurated by the discovery of radioactivity in 1896 and which have enabled us to obtain what seems to be a definitive answer to our question.

The third question is 'How?', or in extended form: 'If the material of which the Earth is composed has been gathered together out of the bounteous supplies in the Universe, what is the cosmic process that can account for the present state of the Earth and—the larger problem—of the Solar System itself?' This is the most difficult problem of all to answer. The method of setting out on the business of finding a solution is one of 'trial and error', that is to say, we state a hypothesis as to the nature of the initial cosmic process and work out its consequence in the light of our knowledge of natural laws; if the result of our investigation does not reproduce the chief characteristics of the Earth and of the Solar System, then the hypothesis must be discarded and we must try again.

With this introduction we now pass on to the detailed arguments in the chapters that follow.

Part I

WHENCE?

CHAPTER II

GENERAL DESCRIPTION OF THE SOLAR SYSTEM

IN this and the two following chapters we shall endeavour to answer the first of the three principal cosmogonic questions, namely, 'Whence came the Earth?'—or, expressed in a slightly different form, 'What is the Earth's parentage?'

Any adequate discussion as to the origin of the Earth according to present-day views involves first of all a general account of the Sun's family of planets and associated bodies (satellites, meteors and comets) which is concisely described as the *Solar System*, for it is the uniformities observable in this system that suggest the clue to the possible elucidation of the problem raised by our question; it may be said at once that, in this connexion, the problem of the Earth's origin is merged in the more comprehensive problem of the origin of the Solar System itself.

It is convenient to begin our survey with a brief reference to the Earth and its relation to the other bodies of the Solar System, leaving an account of the detailed structure of our planet to later pages.

The Earth is almost spherical in form, with a diameter rather less than 8000 miles. Until the time of Copernicus (1473–1543) it was generally believed that the Earth was the centre of the Universe and that, round it, revolved the Sun, the Moon,* the planets then known (Mercury, Venus, Mars, Jupiter and Saturn) and the stars. This geocentric theory of the Universe—in its final form known as the Ptolemaic theory as developed by Ptolemy of Alexandria in the second century A.D.—was a complicated, although ingenious, geometrical

* For photographs of the Moon, Mars, Jupiter and Saturn see Plates I, facing p. 16, II*a*, II*b* and II*c*, facing p. 17.

conception which, however, grew ever more cumbersome and artificial as it strove to keep pace with the increasing accuracy of the observations of heavenly bodies and with the newly discovered phenomena associated with them.

The heliocentric theory of Copernicus introduced ideas of great simplicity; in this theory the Sun was regarded as the central body and around it revolved the Earth and the other planets in circular paths; the Earth was displaced from its proud position as centre of the Universe, in the Ptolemaic theory, and was now relegated to the undistinguished status of a mere planet owing allegiance, like the other known planets, to the Sun as the central controlling body. The familiar rotation of the firmament from east to west was simply explained by Copernicus in terms of the hypothesis that the Earth rotated about an axis in a period of rather less than a day;* we refer to this axis—the diameter joining the Earth's north and south poles—as the *polar diameter* or *rotational axis*. With the invention of the telescope and its subsequent application to astronomical observations by Galileo (1564–1642) in 1609, detailed knowledge of the heavenly bodies increased by leaps and bounds, and the general soundness of the Copernican theory was eventually confirmed beyond the shadow of a doubt. We mention several of the most important and relevant of these early discoveries:

(i) The telescope at once revealed the mountainous nature of the lunar surface (confirming a speculation of Democritus made about a score of centuries before); evidently, the Moon was a heavenly body resembling the Earth in general character and differing from the idealistic object postulated by Aristotle and his successors in the Greek tradition (see Pl. I, facing p. 16).

(ii) The planet Venus was seen to go through phases in much the same way as the Moon, from which it was immedi-

* This is the day according to ordinary usage (more explicitly, a *mean solar day*). The period required by the Earth for a complete rotation is called the *sidereal day*, which is about 3^{m} 56^{s} shorter than the mean solar day.

Lick Observatory

The Moon

PLATE II

Lick Observatory

(*a*) Mars. (*b*) Jupiter. (*c*) Saturn

ately inferred that this planet was not an orb of heaven shining upon the Earth with its own light—as it was supposed, according to pre-Copernican ideas—but a dark body reflecting from its surface the sunlight falling upon it.

(iii) The discovery of sunspots and the study of their apparent motions across the Sun's disk (the latter a somewhat later investigation) showed that the Sun was a body rotating about an axis, thus supporting the Copernican hypothesis as to the Earth's rotation.

(iv) The discovery of the four great satellites (or moons) of Jupiter, which were seen to circulate around Jupiter in much the same way as the Moon circulates around the Earth and as the planets were supposed by Copernicus to circulate around the Sun, brought an argument of irresistible significance in favour of the new doctrine; the satellite system of Jupiter was, in fact, a Copernican solar system in miniature. As telescopic equipment improved, further satellite discoveries were made in due course and the rotation of the planets about their own individual axes was established.

In the Copernican theory the paths, or orbits, of the planets around the Sun were assumed to be circular; the researches of Kepler (1571–1630), based on more accurate observations than those available to Copernicus a century earlier and summarized in his three great laws of planetary motion, showed that each of the planetary orbits was not exactly circular but elliptical, with the Sun situated at a focus of the ellipse. The transcendent work of Newton (1643*–1727) established the law of universal gravitation by which the Sun controlled the planets and maintained their motions in elliptical orbits. And the three laws of Kepler, which were, apparently, completely distinct and unconnected, were now seen to be simple deductions from the great generalization of the Newtonian law.

The discovery of the law of gravitation gave an immense

* According to the unreformed calendar, Newton was born on Christmas Day, 1642.

impetus to the theoretical study of planetary motions; planetary and lunar theory engrossed the attention of many of the most distinguished mathematicians of the eighteenth and nineteenth centuries, the most sensational achievement being the discovery of a new planet, Neptune, in 1846. The planets Mercury, Venus, Mars, Jupiter and Saturn* had been known from time immemorial; Uranus had been discovered in 1781 during one of the numerous and exhaustive sweeps of the heavens made by that indefatigable watcher of the skies, Sir William Herschel (1738–1822).† Before long it was found that Uranus deviated somewhat from the path predicted by the law of gravitation. It is to be remarked that any one planet attracts every other planet according to the law of gravitation and, in calculating the position of Uranus at any time, the effects of the attractions of the other known planets on Uranus were duly taken into account. Using the discrepancies, at various times, between the calculated and observed positions of Uranus, J. C. Adams (1819–1892) and U. J. J. Le Verrier (1811–77), independently and unknown to each other, showed that these discrepancies could be satisfactorily explained on the assumption that they resulted from the attractions on Uranus of an unknown planet beyond the orbit of Uranus; moreover, they were able to specify the unknown planet's position in the sky and in due course on 23 September 1846 the planet, subsequently named Neptune, was seen in the telescope. This was one of the most notable achievements in the history of science and, in particular, it added a new jewel to the crown of Newton's law of gravitation.

The four planets nearest the Sun—Mercury, Venus, Earth and Mars (in order of increasing distance from the Sun)—are small planets, the Earth being the largest of the group with

* For photographs of Mars, Jupiter and Saturn, see Plates II *a*, II *b*, II *c* (facing p. 17).

† Herschel's working rule, 'whatever shines is worth observing', was taken as the motto 'Quicquid nitet notandum' of the Royal Astronomical Society.

a diameter, as already stated, of nearly 8000 miles and Mercury being the smallest with a diameter of about 3000 miles. The next four planets—Jupiter, Saturn, Uranus and Neptune (in order of increasing distance from the Sun)—are very much larger than the planets in the first group, with diameters ranging from about 31,000 miles (for Uranus) to nearly 90,000 miles (for Jupiter). In 1930 a small planet, Pluto, was discovered well beyond the orbit of Neptune; at the present time not very much is known about Pluto, but it is fairly certain that its diameter is not greater than that of the Earth and is probably no larger than that of Mercury. Fig. 2 indicates the relative sizes of the Sun (with a diameter of 865,000 miles) and the nine planets Mercury to Pluto which are known collectively as the *major planets*.

On the first day of 1801 a small planet, later called Ceres, was discovered circulating in a path between the orbits of Mars and Jupiter; this was the first of the *minor planets* to be discovered, about two thousand of which are now known; their orbits all lie effectively between the orbits of Mars and Jupiter. Ceres, the largest of the minor planets, has a diameter of about 500 miles, while the diameter of the smallest so far discovered is hardly likely to exceed 10 miles. It had been surmised by a former generation of astronomers that the minor planets, of which there may be several scores of thousands, are the fragments of a major planet which circulated around the Sun in an elongated orbit lying between the orbits of Mars and Jupiter and which was shattered into fragments at some stage in cosmic history as a result of the powerful disruptive force arising from the gravitational attraction of Jupiter when the hypothetical planet found itself, as a result of special circumstances, in close proximity to Jupiter.

Although the orbits of the planets are elliptical, yet the degree of 'ellipticity', defined in terms of the *eccentricity* of the ellipse, is in most cases so small that for our purposes we can usually ignore it and regard the orbits simply as concentric circles, with the Sun at the centre as in the Copernican plan. Now the time required by a planet to describe its orbit around the Sun (this is called the *orbital period*) can be easily inferred from observations. Also, the third law of Kepler gives very simply the ratio of the distances of any two planets

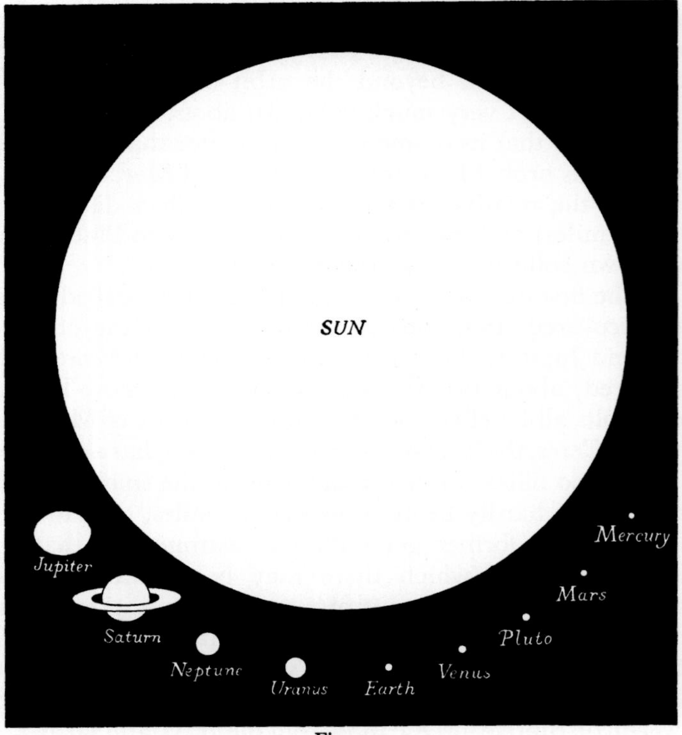

Fig. 2

from the Sun in terms of the ratio of their orbital periods; it is convenient to take one of these planets to be the Earth whose orbital period is *one year* and whose distance from the Sun can be regarded as the *astronomical unit* of distance. It follows, by applying Kepler's law that we can calculate easily a second planet's distance from the Sun in terms of the astronomical unit of distance when once we know its orbital period in years. By this process we can construct a plan, or model, of the Solar System on any convenient scale. As the distance of Pluto from the Sun is rather more than a hundred

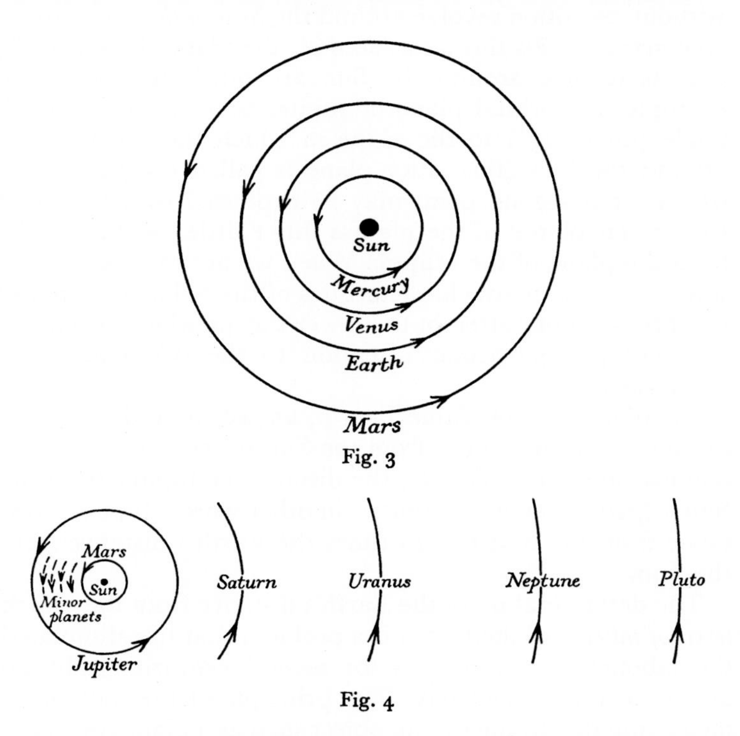

Fig. 3

Fig. 4

times the distance of Mercury from the Sun a single diagram, drawn to scale on a page of this book and representing all the planetary orbits, would be unsatisfactory from the point of view of clarity; consequently, we show the orbits of the first four planets nearest to the Sun in Fig. 3, and parts of the circular orbits of the remaining planets, together with the orbit of Mars, on a much smaller scale in Fig. 4; in this latter figure the reader must imagine the orbits of the vast number of minor planets to be inserted between the orbits of Mars and Jupiter.

There are two important points to notice about the planetary orbits, illustrated in Figs. 3 and 4. First, the planets without exception revolve around the Sun *in the same direction* (the arrows stress this fact); second, the planes in which the planets revolve around the Sun are much the same—for example, the orbital plane of Jupiter is inclined at a small angle (about $1\frac{3}{5}^{\circ}$) to the plane in which the Earth moves around the Sun (this latter plane is called the *plane of the ecliptic*); our second point may be expressed by saying that the orbital planes of the planets differ little, on the whole, from the plane of the ecliptic. When we are in a position to summarize the main characteristics of the Solar System and to embark on our attempt to answer the question 'Whence?' we shall pay particular attention to the two points just mentioned.

In column (2) of Table I on p. 25 we show the average distances* of the planets from the Sun in terms of the astronomical unit; for example, the distance of Jupiter from the Sun is 5·20 astronomical units—in other words, Jupiter's distance from the Sun is 5·20 times the Earth's distance from the Sun.

The determination of the Earth's distance from the Sun *in terms of miles* (or kilometres) is a problem that has stimulated the labours of astronomers for several centuries right up almost to the present day. The principle of the method of measuring the distance of an object such as a planet or a submarine or an aeroplane or a remote mountain-peak is simple; a base-line of known length is essential and from the ends of this base-line observations of the direction of the distant object are made. In Fig. 5 *AB* represents the base-line defined, say, by two vertical poles at a known distance (in

* The average distance refers to the elliptic orbit and, as we are ignoring the ellipticity here, we regard this distance as applying to the radius of the simplified circular orbit.

miles or yards) apart, and *X* is the distant object; the observations consist essentially in measuring the angles *XAB* and *XBA* and since *AB* is known the distance (in miles or yards) of *X* from *A* or from *B* can easily be calculated by the simple process of trigonometry. A range-finder, as used on board a battleship for obtaining, say, the distance of a submarine, is simply an instrument with a base-line, for example of 10 feet, incorporating optical devices for allowing the observer to

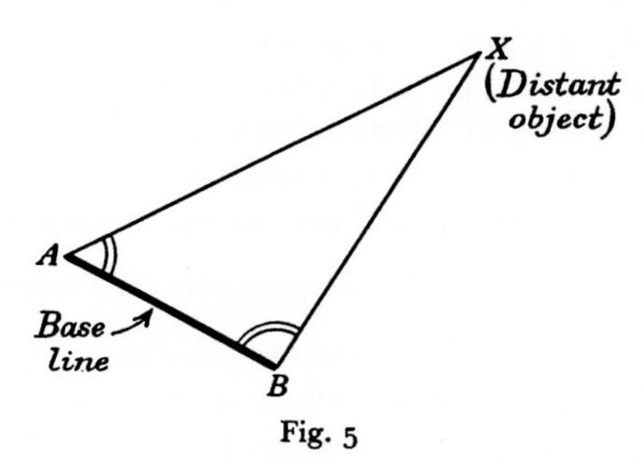

Fig. 5

measure accurately the difference in the direction of the submarine from the ends of the base-line; the appropriate diagram would be drawn with the angles *XAB* and *XBA* equal (see Fig. 5).

The problem of measuring a planetary distance is beset with many difficulties mainly due to the immense distance of the object observed, *X*, from the base-line *AB* which must of necessity be associated with two points whose distance apart is comparable with the dimensions of the Earth. However, let us suppose that, overcoming these difficulties, we succeed in measuring the distance (in miles) of a planet—say, Mars—from the Earth at a given instant. Now, the positions of Mars and the Earth in their orbits at this instant can be calculated accurately and, accordingly, if we make use of the plan of the Solar System as illustrated partially in Fig. 3, we can derive the distance of Mars from the Earth in astronomical units at the instant concerned.* But this distance in astronomical units is, as a result of the actual observations, equivalent to so many miles; hence we readily deduce that the astronomical unit is equivalent to a particular number of miles. For various reasons the most suitable planet for observations in this connexion is not Mars but the minor planet Eros which, under the most favourable circumstances, can approach the Earth to within the comparatively small distance of 14 million miles. In the winter months of 1930–1 observations of Eros were undertaken by a large number of observatories scattered all over the globe and in 1941, after 10 years' work in dealing with the thousands of observations, the Astronomer Royal, Sir Harold Spencer Jones, under whose supervision this immense labour was carried out, announced the final result, namely,

$$1 \text{ astronomical unit} = 93{,}003{,}000 \text{ miles};$$

in other words, the Earth's average distance from the Sun is given by the number of miles just stated.

Knowing now the value of the astronomical unit of distance in miles we can calculate the radii (in miles) of the various planetary orbits from the data in column (2) of Table I; these distances are given in column (3) of this table.

The remaining columns, (4) and (5), of Table I give the orbital periods of the planets and their average orbital speeds in *miles per second*—the latter, for a particular planet, being simply the length of its circular orbit (in miles) divided by the number of seconds in the orbital period. It is to be remarked that the speeds of the planets in their orbits are very much greater than the largest speeds with which we are normally familiar on the Earth; for example, the speed of the fastest jet-propelled aeroplane is rather more than one-fifth of a mile per second (somewhat greater than the speed of sound):

* It need hardly be said that in the actual problem the astronomer has to take into account the fact that the orbits of Mars and the Earth are ellipses and that the orbital plane of Mars is inclined at an angle to the Earth's orbital plane.

TABLE I. *Data Relating to the Orbits of the Planets*

(1) Planet	(2) Average distance from Sun, in astronomical units	(3) Average distance from Sun, in millions of miles	(4) Orbital period	(5) Average speed in orbit (miles per second)
Mercury	0·387	36	88 days	30
Venus	0·723	67	225 days	22
Earth	1·000	93	365¼ days	18½
Mars	1·524	142	1·88 years	15
Jupiter	5·203	484	11·86 years	8
Saturn	9·539	887	29·46 years	6½
Uranus	19·19	1785	84·02 years	4
Neptune	30·07	2797	164·8 years	3⅓
Pluto	39·46	3670	247·7 years	3

the speed of the Earth in its path around the Sun is thus about a hundred times the speed of the fastest aeroplane.*

We summarize in Table II various physical characteristics of the Sun, Moon and the major planets; most of these characteristics are of importance in our later discussions. Let us consider the Earth first. It is found from geodetic measurements that the Earth is not quite spherical but somewhat flattened towards its poles. The polar diameter is 7900 miles while any equatorial diameter (that is, any diameter in the plane of the Earth's equator) is 7927 miles. In a similar way Jupiter and Saturn are considerably flattened; it is almost certain that the other planets share this characteristic, probably to a much less pronounced degree, but the difficulty in making precise measurements of the kind required has precluded any authoritative statement on this point up to the present. In column (2) of Table II the reader will find the diameters of the Sun, Moon and the major planets (where two diameters are given, the first is the polar diameter and the second is the equatorial diameter).

The Sun, Moon and the planets all rotate about axes and a noteworthy characteristic is that the *direction of spin* is, with

* This refers to 1948.

TABLE II. *Physical Characteristics of the Sun, Moon and Planets*

(1)	(2) Diameter in miles	(3) Period of rotation†	(4) Number of satellites	(5) Mass, in terms of the Earth's mass taken as unit	(6) Average density, in terms of the density of water taken as unit
Sun	865,000	25 days	—	333,400	1·41
Moon	2,160	$27\frac{1}{3}$ days	—	$\frac{1}{81}$	3·34
Mercury	3,000	88 days	0	$\frac{1}{27}$	3·73
Venus	7,600	225 days?	0	$\frac{5}{6}$	5·21
Earth	{7,900 7,927	$23^h\ 56^m$	1	1	5·52
Mars	4,200	$24^h\ 37^m$	2	$\frac{1}{9}$	3·94
Jupiter	{82,890 88,800	$9^h\ 50^m$	11	318	1·34
Saturn	{67,200 75,100	$10^h\ 14^m$	9	95	0·69
Uranus	30,900	$10^h\ 48^m$	5	$14\frac{3}{5}$	1·36
Neptune	33,000	$15^h\ 40^m$	2	$17\frac{1}{3}$	1·32
Pluto	Unknown	Unknown	0	Unknown	Unknown

† ^{h} = hour, ^{m} = minute.

one exception, the *same* in all cases;* moreover, this common direction of spin is in the *same sense* † as the direction in which the planets revolve around the Sun. The fact that Jupiter, for example, spins about an axis is inferred from the motion of a marking on its disk and the period of rotation is readily obtained. In the case of planets whose disks are devoid of markings, a spectroscopic method is applied. The rotational periods are shown in column (3) of Table II—the question-mark opposite Venus indicates that the information given in this column is not definitely confirmed. Several minor planets are known to be rotating; this information is deduced from

* The exception is Uranus. We ought also to except Pluto in this generalization since nothing is known at present about its rotation.

† A hypothetical observer, well outside the planetary system and on the northward side of the ecliptic would see the planets revolving around the Sun in a direction opposite to that of the motion of the hands of a clock and the planets spinning about their axes also in this direction.

the fact that their brightness varies in a period of a few hours —the shortest period is 3 hours, that of the minor planet Eunomia—suggesting either that these bodies are irregular in shape or that their reflecting power is not constant over their surfaces.

Column (4) of Table II gives the number of satellites, or moons, which revolve around their parent planets, each satellite system being to some extent a replica on a much smaller scale of the planetary system itself; we shall refer later to these satellite systems in greater detail.

An important physical characteristic of any heavenly body is its mass, defined crudely as the quantity of matter of which it is composed. We are all familiar with such concepts as a pound of sugar, a ton of coal, and so on. It is found by a variety of experiments that the mass of the Earth is about

5000 million million million tons.

The masses of the Sun, Moon and the planets can be deduced by methods which we need not specify; the values are given in column (5) of Table II, in terms of the Earth's mass as unit for convenience: for example, the Sun's mass is 333,400 times the Earth's mass, while the mass of Mercury is not quite 4% of the Earth's mass.

A further physical characteristic of the Sun, Moon and planets which we note here is the average density of these bodies in terms of the density of water as unit (see column (6) of Table II); thus the average density of the Earth is 5·52 times the density of water or, expressed in other words, the mass of the Earth is 5·52 times the mass of a globe of water of the same dimensions as the Earth. A noteworthy feature of column (6) is the much greater density of the four inner and smaller planets (Mercury, Venus, Earth and Mars) as compared with the density of the four outer and giant planets (Jupiter, Saturn, Uranus and Neptune); the density of Pluto, the most remote planet of all, is not known but it is probable that its average density is similar to that of the minor planets.

The Satellites of the Planets

Just as the planets revolve around the Sun in elliptic orbits, so many of the planets have small bodies revolving around them; these are the moons or *satellites*. For example, the body we refer to in ordinary speech as the Moon is the single satellite of the Earth; the Moon describes an elliptic orbit around the Earth in a period of about $27\frac{1}{3}$ days and its average distance from the Earth is about 240,000 miles. The first satellites (other than our Moon) to be discovered were the four large satellites of Jupiter, first seen by Galileo in January 1610 in his newly constructed telescope. These bodies are easily seen in low-power binoculars under normal conditions, being just too faint to be seen by the naked eye; it may be added, however, that several observers have claimed on favourable occasions to have seen the satellites without the aid of a telescope. It would be interesting to speculate on the earlier history of astronomy had the unaided human eye been capable of detecting the four great satellites as they circulated around Jupiter and were carried by the planet in its orbit around the Sun; it is not too much to say that the Ptolemaic, or geocentric, theory of the Universe would have commanded little respect and that astronomical progress would not have been so long retarded by the non-scientific dogmatism of the centuries preceding the telescopic discoveries of Galileo.

The principal details of all the planetary satellites so far discovered are given in Table III. It should be noted that Mercury, Venus and Pluto are without satellites. The names of the satellites are given in column (2), with the exception of the last seven satellites of Jupiter, which have not been specially designated so far and are known simply by their roman numerals. The diameters of those satellites which have been measured are given in column (3); the diameters of the remaining satellites are all small and are probably between 10 and 100 miles. Column (4) gives the average distance of each satellite from the centre of the respective parent planet, and

TABLE III. *The Satellites of the Planets*

(1) Planet	(2) Satellite	(3) Diameter, in miles	(4) Average distance from planet, in miles	(5) Orbital period†
Earth	Moon	2,160	240,000	$27^{d}\ 8^{h}$
Mars	Phobos	About 40	5,800	$7^{h}\ 39^{m}$
	Deimos	About 10	14,600	$30^{h}\ 18^{m}$
Jupiter	Io I	2,110	262,000	$1^{d}\ 18^{h}$
	Europa II	1,865	421,000	$3^{d}\ 13^{h}$
	Ganymede III	3,273	665,000	$7^{d}\ 4^{h}$
	Callisto IV	3,142	1,170,000	$16^{d}\ 17^{h}$
	V	—	112,000	$11^{h}\ 57^{m}$
	VI	—	7,114,000	251 days
	VII	—	7,300,000	260 days
	VIII*	—	14,600,000	739 days
	IX*	—	14,700,000	745 days
	X	—	7,190,000	254 days
	XI*	—	14,000,000	692 days
Saturn	Mimas	370	115,000	$22^{h}\ 37^{m}$
	Enceladus	460	147,000	$1^{d}\ 9^{h}$
	Tethys	750	182,000	$1^{d}\ 21^{h}$
	Dione	900	233,000	$2^{d}\ 18^{h}$
	Rhea	1,150	326,000	$4^{d}\ 12^{h}$
	Titan	3,550	755,000	$15^{d}\ 23^{h}$
	Hyperion	—	914,000	$21^{d}\ 7^{h}$
	Iapetus	—	2,200,000	79 days
	Phoebe*	—	8,003,000	550 days
Uranus	Ariel*	—	118,500	$2^{d}\ 12^{h}$
	Umbriel*	—	165,000	$4^{d}\ 3^{h}$
	Titania*	—	271,000	$8^{d}\ 17^{h}$
	Oberon*	—	362,000	$13^{d}\ 11^{h}$
	Miranda V	—	76,500	31^{h}
Neptune	Triton*	—	218,000	$5^{d}\ 21^{h}$
	Nereid II	—	—	—

* Satellites with retrograde motions (see p. 30).
† ^{d}=day, ^{h}=hour, ^{m}=minute.

the last column gives the times required by the satellites to describe their orbits around their respective planets.

Just as Figs. 3 and 4 give a pictorial representation of the planetary orbits, so in a similar manner we can illustrate the satellite orbits of each planet by an appropriate diagram; we give the diagrams for the satellite systems of Jupiter and Saturn in Figs. 6 and 7.

At first sight the satellite systems of Jupiter and Saturn appear, in general, to be similar to the planetary system except in the matter of scale: the diameter of the planetary system is effectively the diameter of Pluto's orbit around the Sun, namely, about 7300 million miles, whereas the effective

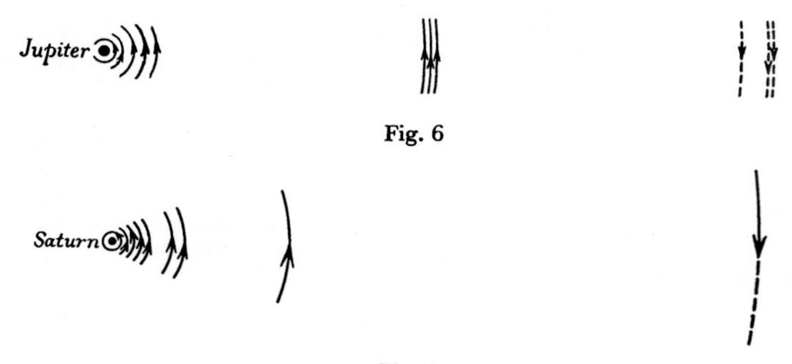

Fig. 6

Fig. 7

diameter of Jupiter's system of satellites is about 29 million miles and the effective diameter of Saturn's system is about 16 million miles. But there is one important difference between the satellite systems of Jupiter and Saturn and the planetary system. We have remarked earlier (p. 22) that the nine major planets and the host of minor planets, without exception, revolve in the *same* direction around the Sun; eight of the satellites of Jupiter (these are the eight nearest to the planet) revolve in the same direction around Jupiter, this direction being that in which the planets revolve around the Sun; the three outermost satellites revolve in the opposite direction and their motion is said to be *retrograde.** Some of the orbits have a comparatively large ellipticity, thus differing

* To draw attention to this feature the orbits of these satellites are shown in Fig. 6 with broken lines.

in this respect from most of the orbits of the major planets; however, in Fig. 6 we ignore this feature and regard the orbits simply as circles. In Fig. 7 the outermost satellite of Saturn is retrograde. The first four satellites of Uranus and the first satellite of Neptune are also retrograde. Of the twenty-eight satellites listed in Table III (omitting Uranus V and Neptune II), the orbits of nineteen are described in what we may call the normal direction, while the orbits of the remaining nine are described in the retrograde direction. The satellites with retrograde orbits are indicated in this table by asterisks.

It is worthy of notice that in Fig. 6 Jupiter's satellites form three distinct groups; the first group consists of five satellites circulating around Jupiter at comparatively small distances, the average distance from Jupiter of the most remote member being about $1\frac{1}{6}$ million miles. The second group consists of three satellites all with average distances from Jupiter of a little over 7 million miles; in the third and most remote group of three satellites the average distances from Jupiter lie between 14 and $14\frac{3}{4}$ million miles. It is this third and most distant group that is abnormal as regards the direction of orbital motion. Similarly, it is the one farthest satellite of Saturn whose orbital motion is retrograde or abnormal. To generalize, we may say that normality is a feature of those satellites of Jupiter and Saturn which are fairly close to their respective planets and are firmly held by the attractive power of their parent planets; on the other hand those satellites which are exceptional or abnormal are at the farthest boundaries of the satellite systems and are subject to the comparatively feeble gravitational attraction of their controlling planets. We shall have occasion later to discuss these features of the two great satellite systems.

METEORS

Everyone is familiar with meteors—or shooting-stars, as they are popularly called. It is not always realized how small, in general, is the cosmic body responsible for the bright but quickly evanescent trail of light across the night sky. An ordinary meteor may be no larger than a pebble or even a grain of sand; its startling effect is a consequence of its immense speed as it rushes into the Earth's atmosphere. This speed is generally from about 20–40 miles per second and the friction on the meteor resulting from its rapid passage through the air causes the solid material to be vaporized, the temperature being so great that the vapour and the air in the immediate neighbourhood are rendered highly luminous. It might be thought that such a small body as a meteor could be of very little scientific interest, but this is actually far from being the case. Careful observations of the track of a meteor, as viewed from two or more positions on the Earth's surface, enable us to calculate the height above the Earth's surface at which the meteor becomes luminous and this information evidently gives us a clue to the thickness of the atmospheric layer by which the solid Earth is enveloped.

As a result of many thousands of observations of this type it is found that the atmosphere extends to a height of 100 miles at least, the density of the air at this height being just adequate to bring into play the frictional resistance necessary to cause luminescence. But the atmosphere must extend to still greater heights. In recent years a variety of methods for exploring the upper atmosphere—including the study of sound-waves generated by explosions, of radio-waves and of the aurora borealis—give a fairly complete picture of the physical characteristics of our envelope of air; it is concluded that even up to a height of 500 or 600 miles there are still traces of atmosphere, of such a low density, however, as to be little better than a vacuum.

It is pertinent to our main line of investigation to inquire into the origin of meteors. Are they members, however insignificant, of the Solar System in the sense in which we regard the planets as members, or are they visitors from extra-planetary space speeding through the void during countless ages until, in the fullness of time, the circumstances of their motion lead them into our upper atmosphere where they perish in a streak of glory? From the study of the paths of meteors it is inferred that, although possibly some of these bodies do actually arrive from outer space, the great majority must be reckoned to be as closely associated with the Solar System as the planets themselves. Certainly there is no doubt about the character of the great meteor showers which occur on or around certain dates during the year. For example, the display of shooting-stars, called the *Leonids*, between 13 and 16 November is due to a vast swarm of meteors pursuing an elongated path around the Sun in a period of $33\frac{1}{3}$ years; on or about 14 November the Earth in its path around the Sun crosses the track of the meteors and nets in its atmosphere such meteors as happen to be in that part of the orbit. Once in 33 years the Earth encounters the main swarm and then a display of unimagined grandeur is observed when thousands of meteors flash across the sky in the space of a few hours. Such was the display of the Leonids in 1866 and on various other recorded occasions; however, in 1899 and 1932 the displays fell short of expectation, possibly because the Earth just failed to make full contact with the main swarm or possibly because the number of meteors, still pursuing their paths around the Sun, was very markedly reduced as a consequence of previous encounters with the Earth.

METEORITES

So far in dealing with meteors we have been concerned with bodies of small size which are vaporized before they can reach the Earth's surface. Occasionally, a piece of cosmic matter of very much greater size than the ordinary meteor hurtles into our atmosphere and succeeds in penetrating to the Earth's surface. Such a body is called a *meteorite*, or aerolite, to distinguish it from the meteor whose existence is only revealed by its bright streak in the sky. Meteorites weighing anything between a few pounds and a few tons and bearing all the signs of their atmospheric passage have been picked up in various parts of the Earth and now, their celestial wanderings over, they repose in the quiet seclusion of our museums.*

Many meteorites have left their imprint on the Earth in no uncertain way; it will suffice to mention only two instances; first, the great Siberian meteorite, which in 1908 devastated an area of several hundred square miles—fortunately in a sparsely populated part of northern Siberia—, and second, the meteorite, still uncovered, which produced the large Arizona crater, about three-quarters of a mile in diameter and several hundred feet deep. Whether meteorites are classed as bodies forming part of the Solar System or, alternatively, as visitors from extra-planetary space, they provide evidence of the utmost value in cosmogonic speculations; what this evidence is we shall discuss in sufficient detail in a later part of the book.

Comets

There is one more type of celestial body associated with the solar system which requires mention; this is the *comet*. One of the most beautiful photographs of comets—that of Morehouse's Comet which appeared in 1908—is shown in Plate III (facing p. 52). Several comets are periodic: for example, Halley's Comet (the appearance of which in 1066 is pictorially represented in the famous Bayeux tapestry) pursues a very elongated path around the Sun in a period of about 76 years; another well-known comet, Encke's Comet, which is, however, a rather faint object, has a period of about $3\frac{1}{3}$ years. We have remarked earlier that the orbits of the planets, although ellipses, differ not very notably from circles; cometary orbits, on the other hand, are extremely elongated. For example, at its nearest approach to the Sun Halley's Comet, which last appeared in 1910, comes within about

* For example, one of the largest known meteorites is that found in Greenland by the late Admiral Peary, U.S.N.; it weighs about 26 tons and is now in the American Museum of Natural History, New York. The largest, weighing about 60 tons, was found near Grootfontein in South West Africa.

14 million miles of our luminary, while at the most remote point of its orbit it is more than 3000 million miles away. A second point of difference between cometary and planetary orbits is that, whereas all the planets revolve around the Sun in the same direction, there is no such uniformity in the directions of the orbital motions of the comets. A further point of difference may be noted: the planes of the planetary orbits are, on the whole, close to the plane of the ecliptic; this is not true of the planes of the cometary orbits.

The distinguishing feature of a bright comet is its tail (or tails, as in the case of Morehouse's Comet) which, originating in the 'head' of the comet, may stretch for scores of millions of miles across the sky. It is highly probable that the head of a comet consists of a very considerable aggregation of matter varying in size from the finest dust particles to solid chunks as large as meteorites—or perhaps occasionally even as large as the smallest minor planets or satellites—together with solidified or liquefied gases of several kinds. When the comet approaches the Sun, the intense radiation of the Sun exerts a repelling effect, which is known as the *pressure of radiation.* It acts on the minute particles of dust and the molecules of gas (now vaporized from the solid or liquid state by the Sun's heat), and the result is the formation of the spectacular tail. Owing to the comparatively feeble gravitational power of the cometary head, most of the tail is eventually dissipated into space. Not all comets, however, are observed to have tails and it is inferred that after perhaps hundreds or thousands of approaches to the Sun the tail-forming material has been wholly diffused into the voids of space, leaving the comet merely as an undistinguished collection of meteoric matter loosely held together by its feeble gravitational attraction. We have already referred to the shooting-stars known as the Leonids; this swarm of meteors pursues an orbit identical with that of Tempel's Comet and it is a fair inference that the Leonids are merely those meteors which have gradually escaped from the weak gravitational control of the main mass of the comet.

It may be that several comets are permanent members of the Solar System in the sense applied to the planets; on the other hand it seems more than probable that some comets have reached the environs of the Solar System from extra-planetary space and, coming within the gravitational attraction of the Sun and planets, have been 'captured', thereafter describing elliptical and highly elongated orbits around the Sun. The reverse process is also known. For example, before its close approach to the Sun and while it was still hundreds of millions of miles away, Comet Morehouse moved in an orbit that was deduced from careful calculations to be elliptical. In other words, the comet would be defined as belonging to the family of periodic comets. But, in its nearer approach to the Sun, the comet came within the gravitational influence of one or more of the planets with the result that the characteristics of its orbit were altered in such a way that, in sweeping round the Sun, the path became 'hyperbolic'. By this time (1950) the comet must have reached the outermost bounds of the Solar System and, still heading away from the Sun, it will disappear into the emptiness of space, never to return to the bosom of the solar family.

Referring to p. 34, where the principal differences between cometary and planetary orbits are mentioned, we see that the cometary orbits show none of the uniformities associated with the planetary orbits. It would seem to follow that, anticipating the answer to the first of our three questions, we cannot accord to the comets, as a whole, the distinction of being original members of the Solar System. It has recently been suggested by Bobrovnikoff that these wayward bodies were acquired by capture when, it is assumed, the Sun passed through a cloud of diffuse matter, perhaps a million years ago. This suggestion is, of course, highly speculative but something of the kind is necessary to account in some measure for the presence of comets within the boundaries of the comparatively staid planetary system.

CHAPTER III

GENERAL DESCRIPTION OF THE EARTH

In this and the following chapter we shall describe the main physical characteristics of the principal members of the Solar System. If, as has already been suggested (p. 10), the planets and satellites have a common origin, we might expect to see one or more features shared by them, at least by those bodies for which observational evidence is adequate; further, if there are striking dissimilarities, we might hope to be able to explain these in terms of physical processes acting through the long succession of centuries since the occurrence of the event that gave birth to the Sun's family of planets and satellites. For example, the mountainous character of the surfaces of both the Earth and the Moon is a point of similarity that is immediately evident to anyone who has viewed the Moon through a telescope or has seen a photograph of the lunar surface such as that shown in Plate I (facing p. 16). As an example of dissimilarity between the Earth and the Moon we note the complete absence of an atmosphere on the latter,* whereas the existence of the terrestrial atmosphere is a matter of universal knowledge and is indeed of vital necessity for nearly all forms of organic existence. We shall see later that, although it cannot be proved that the Moon actually possessed an atmosphere at any time in the remote past, yet if the existence of a lunar atmosphere in the past is granted then there are very sound and conclusive arguments to account for its absence at the present time.

Our survey begins most profitably with the Earth; however, in this chapter, we restrict our account to features

* This statement of fact will be amplified in due course.

of a rather general character, leaving matters of detailed geological interest to a later stage. As our discussion involves a certain amount of attention to the properties of light, we take the opportunity of including an account of this topic sufficient for our purpose: this inevitably leads us to discuss the chemical constitution of the Sun.

As we have seen, the Earth is a nearly spherical body, flattened slightly at the poles. The principal scene of our activities is the outer surface of what is called the Earth's solid crust, with the oceans, seas and lakes filling many of the depressions, and mountains and hills forming protuberances. The solidity of the crust is in contrast with the possible fluidity of the core as suggested by the molten state of the lava ejected by active volcanoes from the Earth's inner recesses. Further, we draw the conclusion that the Earth's interior is hotter than the outer crust, from evidence furnished by hot springs, the ejection of steam, the high temperature of molten lava (about 1000° to 1200° C.) and the observed increase of temperature with depth in mines and deep borings. Surrounding the Earth is the terrestrial atmosphere extending to a height of several hundreds of miles and consisting mainly of oxygen and nitrogen in the proportion of 1 to 4 approximately, together with very much smaller or even minute proportions of water-vapour, carbon dioxide, hydrogen and the chemically inert gases helium, argon, neon (familiar in the luminous advertising signs), krypton and xenon, of which argon is the most abundant forming about 1% of the atmosphere.

Thus in our first rapid view of the Earth we see a body, nearly spherical in form, with an outer solid skin or crust overlying a possibly fluid core, the temperature increasing with depth below the surface, and the whole sphere surrounded by an envelope of gases extending to a great height.

Let us first consider the atmosphere. A question that immediately suggests itself to the reader is the following: Are there any special circumstances why the Earth should have an atmosphere while the Moon and all but a few of the members of the planetary system have none? Our knowledge of the characteristics of gases supplies a definite answer.

A gas such as hydrogen or oxygen under ordinary terrestrial conditions consists of molecules, a molecule being the smallest particle which can retain the usual familiar properties of the element concerned; this definition of molecule applies to any single substance, element or compound, whether gaseous, liquid or solid. Thus, in the formation of water from hydrogen and oxygen, two molecules of hydrogen combine chemically with one molecule of oxygen to produce two molecules of water. A more fundamental physical entity is, however, the atom and a single molecule of such a gas as hydrogen or oxygen under ordinary conditions is a firmly knit combination of two atoms; such gases are called *diatomic*. The chemically inert gases helium, argon, neon, krypton and xenon are *monatomic*, that is to say, the molecules are identical with atoms. Gases such as water-vapour and carbon dioxide are *polyatomic*—the molecule consists of more than two atoms of the constituent elements.

Gaseous molecules, and molecules in general, are extremely minute structures and if we picture them as tiny spheres and suppose them laid in a straight line in contact, like golf balls in the cylindrical receptacle on the starting tee of a golf course, there would be from 50 millions to 100 millions of molecules in a single inch, the actual number depending on the nature of the gas. In a gas the molecules are not in contact and must be imagined to be rushing about in all sorts of directions with all sorts of speeds and colliding with other molecules, the number of collisions suffered by any one molecule in one second being several thousand millions. If the gas is contained in a vessel, just as air is contained in a bicycle tyre, the molecules are also continually impinging on the inner surface of the containing vessel, this ceaseless bombardment giving rise to what we call the pressure of the gas; the greater the number of molecules within the vessel the greater is the pressure so that, to increase the pressure in a tyre we pump more air into it thereby rendering it more efficient in supporting the weight of the rider. Also the pressure increases with the temperature of the gas for, as the temperature is raised, the activities of the molecules are increased so that their impacts on the containing surface become more powerful. This conception of a gas has led to the mathematical development known as the kinetic theory of gases in which the properties and behaviour of a gas are studied by statistical methods; like any other theory of repute, the kinetic theory is, in the main, satisfactorily supported by observations and experiments.

In a gas, such as the oxygen or nitrogen of the air, the molecules are anything but tightly packed together: they resemble in this respect the twenty-two players in an Association football match who are free to roam over an area of 120 by 60 yards. By the kinetic theory it can be calculated that at the ordinary temperature the average distance traversed between collisions by a molecule of atmospheric oxygen is about $\frac{1}{160,000}$ of an inch. This is known as the mean free path; it depends on the nature of the gas and its temperature and pressure and is analogous, in our illustration, to the average distance between consecutive 'collisions' of our football players. In a gas, then, the molecules have considerable freedom of movement. In a liquid, on the other hand, the molecules are much more closely packed together, resembling the spectators on the well-filled, but not tightly packed, terraces of a football ground where freedom of movement, although not easy, is yet practicable. In a solid the molecules are jammed tightly together like the spectators at an international match or 'cup final', the possibility of movement being then extremely restricted.

Molecular movements in a liquid can be conveniently illustrated with reference to what is usually described as the 'Brownian movement', so called after Robert Brown, an English botanist, who first studied the phenomenon in 1827; the true explanation was given only half a century later.

Observing with a microscope, Brown noticed that minute particles of pollen, suspended in water, appeared to be endowed with life, so far, at least, as activity was concerned; for they darted hither and thither, pursuing zigzag paths apparently of a completely random character, like minnows in a pool. Minute particles of other substances exhibited the same phenomenon. Although fairly well packed, the molecules of water are continually in motion and continually colliding with neighbouring molecules and with anything, such as pollen, in their paths. A particular speck of pollen is thus being bombarded on all sides, but as it is very minute the number of bombarding molecules is comparatively small. As the molecular velocities are not all the same in magnitude and direction, it is easy to see that the effect of the upward molecular impulses on the pollen at a given instant may be greater than the effect of the downward impulses, in which case the pollen will move in an upward direction for an instant. The excess of molecular impulses may, of course, occur in any direction at subsequent instants and, consequently, the particle of pollen will describe a zigzag path that appears to be completely unpredictable.

The molecules of different gases, although very much of the same size as already indicated, vary considerably in weight; the *molecular weights* of several gases are shown in column (2) of Table IV (p. 43) on the basis, which at first sight appears rather peculiar, that the molecular weight of oxygen is denoted by 32·0. The molecular weights in the table refer really to the *ratios* of the actual weights of the molecules of the various gases; for example, the ratio of the weights of the molecules of hydrogen and oxygen is 2·016:32·0, in other words, the molecular weight of oxygen is approximately sixteen times the molecular weight of hydrogen. The actual weights of the molecules of the elements are exceedingly minute but they can be determined: for example, the weight of a molecule of hydrogen is such that about 136 million million million million molecules would have to be lumped together to give a weight of one pound; the corresponding number of oxygen molecules in a weight of one pound is about 8½ million million million million.

We have seen that the dimensions of molecules are also exceedingly minute. If we suppose that the molecules in one gramme of oxygen are placed in contact in a straight line, they would extend to a distance of about 4300 million miles, that is, about forty-six times the distance of the Sun from the Earth. We should require only about 13 pounds of oxygen for the molecules to extend from the Earth to the nearest star (this distance is about 25 million million miles).

Again, the average velocity of the molecules of a gas can be calculated; this velocity depends on the temperature of the gas, becoming greater as the temperature increases. The average molecular velocity plays an important role in the study of atmospheres. In Table IV we give in column (3) the average molecular velocities, in *miles per second*, of the gases listed, for the temperature 0° C. (the temperature of freezing water).* The remaining columns give the average molecular velocities corresponding to several temperatures ranging from −200° C. to 5000° C. We shall have occasion to refer frequently to the information contained in this table.

It is to be remembered that the molecular velocities in Table IV are average values and that within any volume of gas, in addition to molecules moving with less than the average velocity, there will be a proportion of molecules with velocities twice, thrice, ... the average velocity; however, this proportion diminishes very rapidly as the velocity concerned increases beyond the average velocity; thus, only a very minute proportion of the molecules will have velocities which are, say, five times the average velocity.

The average molecular velocity of a given gas is particularly relevant in any discussion of the capacity of a planet or satellite to retain the gas as part of its atmosphere. When

* C. refers to the Centigrade scale of temperature.

TABLE IV. *Molecular Weights, and Average Molecular Velocities for Different Temperatures*

(1)	(2)	Average molecular velocities (miles per second)					
	Molecular weight (oxygen)	(3)	(4)	(5)	(6)	(7)	(8)
Gas	32·0	0° C.	200° C.	500° C.	2000° C.	5000° C.	−200° C.
Hydrogen	2·016	1·14	1·50	1·93	3·30	5·03	0·59
Helium	4·002	0·81	1·07	1·36	2·34	3·57	0·42
Nitrogen	28·02	0·31	0·40	0·52	0·88	1·34	0·16
Oxygen	32·00	0·29	0·38	0·48	0·83	1·26	0·15
Neon	20·18	0·36	0·48	0·61	1·04	1·59	0·19
Argon	39·94	0·26	0·34	0·43	0·74	1·13	0·13
Krypton	82·17	0·18	0·23	0·30	0·51	0·78	0·09
Xenon	131·3	0·14	0·19	0·24	0·41	0·62	0·07
Water-vapour	18·02	0·38	0·50	0·64	1·10*	1·68*	0·20
Carbon dioxide	44·01	0·24	0·32	0·41	0·70*	1·07*	0·13
Methane	16·04	0·41	0·53	0·68	1·17*	1·78*	0·21
Ammonia	17·03	0·39	0·52	0·66	1·14*	1·73*	0·20

* These, and also the entries in column (8), are theoretical values calculated on the assumption that the substances concerned could exist in the gaseous form at the corresponding temperatures.

a bullet is fired vertically upward, its motion will be retarded owing to the gravitational attraction of the Earth (for simplicity, we disregard the retarding influence of the atmosphere) until it comes momentarily to rest; it then begins its downward fall, eventually returning to the Earth's surface. The height attained by the bullet depends on its initial velocity; if the velocity is doubled, the height is substantially four times greater, and so on, provided that the muzzle velocities with which we are dealing are a few hundred feet per second. Suppose that the muzzle velocity is 10 miles per second; what, then, happens to the bullet? In this case calculation shows that the bullet would pass beyond the Earth's gravitational control and would escape from the Earth, pursuing its path in interplanetary space and thereafter, probably, in interstellar space. The minimum velocity of projection at the Earth's surface to ensure that the bullet will escape for ever from the Earth, the effect of the atmosphere retarding the bullet being ignored, is called the *velocity of escape*; for the Earth it is 7 miles per second.

Referring to column (3) of Table IV, corresponding to the temperature 0° C. for the gases listed, we see that the velocity of escape from the Earth is about six times the average molecular velocity of hydrogen, about nine times the average molecular velocity of helium, and still greater for the remaining gases. At the highest levels of our atmosphere the air is so rarefied that a molecule of hydrogen, for example, might quite well avoid collisions with the comparatively few molecules in its neighbourhood and escape into interplanetary space provided its outward velocity exceeded 7 miles per second.* This is the normal method by which any planetary atmosphere can be eventually dissipated into space; the rapidity of the process depends on the proportion of molecular velocities equal to or greater than the velocity of escape; if, on the other hand, this proportion is negligible, the gases of the atmosphere are held firmly by the planet's gravitational attraction.

Since the average molecular velocity of any gas increases with temperature, as illustrated in Table IV, we require some information as to the present and past temperatures of the planetary atmospheres. It may be said at once that the present temperatures are known with sufficient accuracy, at least, to enable us to apply the principles of the kinetic theory with confidence. As to past temperatures we have to recognize that, over a long period of time, the Earth and the other planets are cooling, an inevitable and unescapable feature

* The velocity of escape has been defined with regard to projection from the surface of the Earth; if we consider a molecule at a height of 540 miles, say, above the Earth's surface, the velocity of escape from this level is about 6½ miles per second. In the sequel we shall generally ignore the differences introduced in this way.

in their physical evolution; in the very distant past, their temperatures must have been very much higher and it is legitimate to consider the possibility that at one time the planets were molten or even gaseous, with temperatures of a few thousand degrees; to take account of this possibility Table IV contains the average molecular velocities for temperatures of 2000° and 5000° C.

In considering more precisely the capacity or the incapacity of a planet to retain an atmosphere whatever the temperature may be, we make use of the calculations of Sir James Jeans in this connexion. If the average molecular velocity of a particular gas is one-quarter of the velocity of escape, the proportion of molecules with actual outward velocities equal to, or exceeding, the velocity of escape is not inconsiderable: these molecules will accordingly escape from the gravitational control of the planet concerned, the rate being such that the atmosphere, so far as the particular gas is concerned, will be dissipated completely in about 50,000 years; if the average molecular velocity is greater than one-quarter of the velocity of escape, the atmosphere will, of course, disappear more rapidly. If the average molecular velocity is two-ninths of the velocity of escape, the gas will be dissipated in about 30 million years. If the average molecular velocity is one-fifth of the velocity of escape, the atmosphere will be lost in about 25,000 million years and more slowly still, of course, if the average molecular velocity is less than one-fifth of the velocity of escape. As we shall not be contemplating periods of past time exceeding several thousand million years we can take as the two critical values of the average molecular velocity: (i) one-quarter of the velocity of escape, (ii) one-fifth of the velocity of escape. In the case of (i) the atmosphere will disappear almost instantaneously, in relation to the large time-interval of several thousand million years under contemplation; while in the case of (ii) we can say that the atmosphere will be substantially retained. To anticipate later conclusions we regard the large time-interval mentioned—more precisely,

3000 or 4000 million years—as the period during which the Earth and the other planets have existed as independent bodies, beginning their careers as intensely hot molten globes with temperatures of several thousand degrees; we must suppose them to cool very rapidly at first and eventually, after this long interval of time, to attain the physical states in which we see them at present.

The Earth's Atmosphere

We now apply these principles to the Earth's atmosphere. Remembering that the velocity of escape from the Earth is 7 miles per second we see that the two critical average molecular velocities are: (i) one-quarter of 7 miles per second, that is, $1\frac{3}{4}$ miles per second, and (ii) one-fifth of 7 miles per second, that is, $1\frac{2}{5}$ miles per second. If the average molecular velocity of a terrestrial gas exceeds $1\frac{3}{4}$ miles per second, the gas will disappear into the voids of space in a very short time—almost instantaneously, on the cosmic time-scale. Whereas, if the average molecular velocity is not greater than $1\frac{2}{5}$ miles per second, the gas will be effectively retained throughout the lifetime of the Earth as an independent planet.

We consider hydrogen first. As already stated, this gas exists only in very small quantities in the terrestrial atmosphere and, even allowing for the amounts of hydrogen in combination with oxygen as water and in other chemical compounds of the Earth's crust, we must conclude that hydrogen is very poorly represented amongst the elements constituting the Earth. This, at first sight, is surprising when we consider the preponderating abundance of hydrogen in the Universe, for recent research is confident in affirming that hydrogen constitutes about one-third of the Sun and many of the stars; and it is not improbable that this proportion applies in rough measure to the whole material universe. There is no valid reason for thinking that the Earth failed to have its proper ration of hydrogen when it came into existence as a hot globe. How, then, did the hydrogen disappear? The data in Table IV (p. 43) supply a convincing answer. As we have seen the critical average molecular velocities for the Earth are $1\frac{3}{4}$ miles per second, corresponding to almost immediate dissipation, and $1\frac{2}{5}$ miles per second, corresponding to the virtually complete retention of the gas. For hydrogen, the table shows that an average molecular velocity of $1\frac{3}{4}$ miles per second corresponds to a temperature of about 400° C., so that, if the atmosphere had a temperature greater than this, the hydrogen could not be retained. Considering the very much higher temperatures prevailing in the earlier stage of the Earth's existence, we have now very little difficulty in accounting for the almost complete disappearance of hydrogen from the terrestrial atmosphere. When the atmosphere had cooled to about 150° C., to which the second critical velocity of $1\frac{2}{5}$ miles per second corresponds, such hydrogen as remained would be substantially retained, provided that no other physical process, not yet envisaged, came into operation; we shall deal with this point when we consider the atmospherical abundance of helium.

When we apply the same arguments to the other gaseous constituents of the atmosphere, the columns for the temperatures 0°, 200° and 500° C. show that the average molecular velocities of these gases are below the second critical velocity of $1\frac{2}{5}$ miles per second, although helium is just on the border-line for 500° C.; it follows that these gases should be easily retained—some, of course, more easily than others. If we go further back in the Earth's history when, if ever, the temperature was 5000° C. we see from the appropriate column that the elements argon, krypton and xenon would be easily retained despite this high temperature, and that nitrogen and oxygen, being just a little below the border-line, would be substantially at least, if not wholly, retained; helium and neon would undoubtedly be dissipated, the former almost completely and the latter very substantially, because the average molecular velocity of neon at this temperature is 1·59 miles per second, intermediate between the two critical velocities of $1\frac{2}{5}$ and $1\frac{3}{4}$ miles per second. Although, in the Universe, neon is not a rare gas the scarcity of neon in the terrestrial atmosphere is thus satisfactorily explained on the assumption that the Earth's temperature was several thousand degrees in the distant past. But we cannot explain the almost complete absence of helium from the terrestrial atmosphere *at present* in this way, although it is still correct to affirm that the *original* amount of atmospheric helium would be quickly lost from the very hot primitive Earth.

In our arguments concerning the dissipation or retention of terrestrial gases it has been tacitly assumed that there has been no replenishment of any particular gas by means of physical or chemical processes operating within the Earth. This assumption does not hold in the case of helium, for this gas is a product of the spontaneous disintegration of radioactive elements, such as uranium, radium and thorium, and although it may be temporarily imprisoned within the rocks containing the radioactive elements, it eventually finds its way into the atmosphere owing to the processes of erosion or to volcanic action. The amount of helium added to the atmosphere in this way has been estimated by geologists and it is found to be very many times greater than the amount actually present in the atmosphere to-day. If we go no further back than the time when the atmospheric temperature was 500° C., we see from Table IV that the helium produced, since then, by these radioactive processes should have been firmly retained in the atmosphere; the analysis of the composition of our atmosphere shows that all but a minute fraction of the helium has disappeared. What has happened to the helium? Being an inert gas it cannot form chemical compounds and so cannot disappear from the atmosphere in this way; there must be a satisfactory explanation of a different kind. The explanation is based on the peculiar properties of atoms; it will be convenient to postpone further discussion till a little later (p. 55) when we deal with the phenomenon of atomic radiations at greater length.

Oxygen is another gas whose atmospheric content is liable to change; it enters into the chemical actions of combustion (the formation of carbon dioxide, in particular), the respiration of animals, the rusting of iron and so on; it is returned to the atmosphere as a result of the dissociation of carbon dioxide in the respiratory action of many forms of plant life. Consequently, it is unlikely that the amount of oxygen in the atmosphere has remained constant for millions of years, although at the present time there may be a rough balance between the abstraction of oxygen from the atmosphere and its restoration. It may be added that, for the reasons indicated, the atmospheric content of carbon dioxide has almost certainly varied, perhaps not inconsiderably, during the Earth's lifetime.

To summarize, we can conclude that under conditions of terrestrial temperatures not markedly different from those prevailing at present the Earth's atmosphere must have existed for a very long time and will be retained into the distant future, although its actual composition may be subject to variations.

The Ozone Layer

There is one feature of the Earth's atmosphere which is, literally, of vital importance to all forms of terrestrial life; this is the ozone layer. We have seen that oxygen is one of the principal constituents of the atmosphere and that its molecules are diatomic, that is to say each molecule consists of two atoms firmly bound together. But it is possible for oxygen to be triatomic, three atoms forming a single molecule, and in this form the oxygen is called *ozone*. Although the atmospheric content of ozone is extremely small—it is mainly from 10 to 25 miles above the Earth's surface—yet its influence is out of all proportion to the quantity involved. To understand its peculiar function we must consider for a little the simpler characteristics of light.

The Properties of Light

We are all familiar with the system of waves produced when a small pebble is dropped into the unruffled water of a pond; the distance between two consecutive crests of the waves is called the *wave-length*. The energy of the falling pebble has been mostly transformed into the energy of wave-motion and the actual wave-length observed on any occasion will be dependent on the amount of energy transmitted to the water: on a pond the wave-length may be an inch or two; in the Atlantic, with more powerful disturbing factors operating, the wave-length may be several hundred feet. The essential feature of wave-motion is the propagation of energy from the point of the initial disturbance—taking the case of the pond—to distant parts. Now, light is a form of energy and, according to the wave-theory of light, it is propagated in waves from the source of the initial disturbance located in the individual atoms or molecules of matter. For example, the atoms in the filament of an electric-light bulb are continuously supplied with energy derived from the electric current and, as continuously, the atoms part with this energy in the form of waves of light (and heat) radiated in all directions from the filament. We usually designate by *light* that particular kind of energy to which the eye is sensitive. But, as we shall see, the wave-lengths of light are confined within a narrow range and it is convenient to use the term *radiation* to denote all energy of this kind whatever the wave-length may be.

The principal radiations in order of increasing wave-length are: gamma-rays, X-rays, ultra-violet radiations, visible light, infra-red, short electric waves and radio waves (Fig. 8). The physicist has developed accurate laboratory methods for the measurement of wave-lengths; except for radio waves which have wave-lengths varying from a few feet to several thousand feet, the wave-lengths of the other radiations mentioned are extremely small and it is found convenient to use a special unit of length with which to refer to them. This unit is called the

Ångström unit, denoted by A., and is such that a hundred million of these units go to make up a centimetre (about two-fifths of an inch). For example, the red light emitted by glowing cadmium vapour is found by laboratory experiments conducted under certain specified conditions to have a wave-length of 6438·4696 A., that is, about 64 millionths of a centimetre.

Sunlight is found to consist of all wave-lengths from about 4000 to about 8000 A. The familiar rainbow is Nature's experiment for demonstrating the composite character of

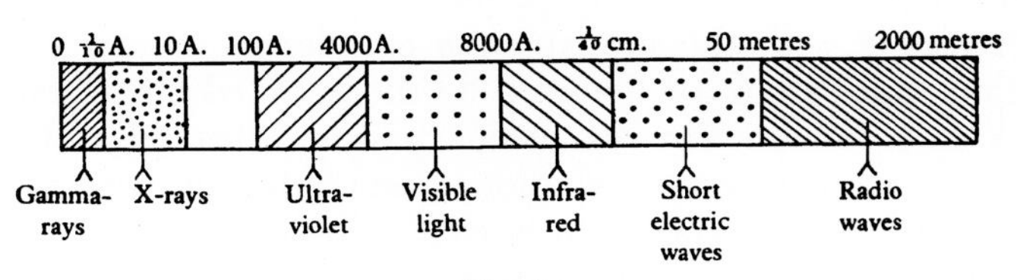

Fig. 8

sunlight; raindrops break up the sunlight into the succession of the principal colours violet, indigo, blue, green, yellow, orange and red—this is Nature's *spectrum*. The colour red is associated with the longest wave-lengths of the spectrum between about 6000 and 8000 A., merging at the lower wave-length into the colour orange; at the other end of the succession of colours, violet is associated with wave-lengths between 4000 and 4500 A. approximately, merging into the colour indigo at the larger wave-length; similarly, intermediate colours are associated with particular ranges of wave-lengths. The spectrum colours are shown in Plate IV*a* (facing p. 53). Radiations with wave-lengths from about a hundred Ångström units up to about 4000 A. (the beginning of the visible spectrum) are called *ultra-violet* radiations and the radiations with wave-lengths from about 8000 A. to several thousand Ångström units beyond are known as *infra-red* radiations; beyond these again (and including the infra-red) are the radiations to which we commonly apply the term 'heat'; still farther on are short electric waves and radio waves (Fig. 8).

The Sun's Spectrum

A glass prism performs the same function as the raindrops in breaking up sunlight into its constituent wave-lengths and the instrument used for producing the spectrum is called a spectroscope—or spectrograph, when the spectrum is photographed. Accordingly, when we view sunlight that has been broken up by the prism of the spectroscope we see in all its beauty the spectrum of sunlight with its range of rainbow colours (Plate IV *a*, facing p. 53). But the combination of a telescope with the spectroscope enables us to see certain details that cannot be detected by the unaided eye when we look at a rainbow in the ordinary way. Instead of being a continuous succession of colours, one fading imperceptibly into the next, the spectrum is seen to be crossed by a large number of dark lines varying in width and in the intensity of 'darkness'; these lines are called *absorption lines* and the spectrum, consisting of the rainbow colours and the absorption lines is called an *absorption spectrum.* Some of the principal absorption lines in the solar spectrum are shown in Plate IV *b*, (facing p. 53).

As the interpretation of such a spectrum is based on laboratory investigations, we shall briefly describe three experiments illustrating the principal characteristics with which we are concerned.

Experiment 1. If a small amount of any compound of sodium (common salt, for example) is held in a Bunsen flame, the latter is coloured a brilliant yellow; when this yellow light is examined in a spectroscope, it is seen to consist of two bright yellow lines close together, whose wave-lengths are 5890 and 5896 A. (known as the *D* lines of sodium); these bright lines are called *emission lines* and are characteristic of the element sodium. We infer that, under the conditions of the experiment, the sodium atoms possess the property of emitting light of the two wave-lengths mentioned. The emission of light implies the discharge of energy and this energy is acquired by the

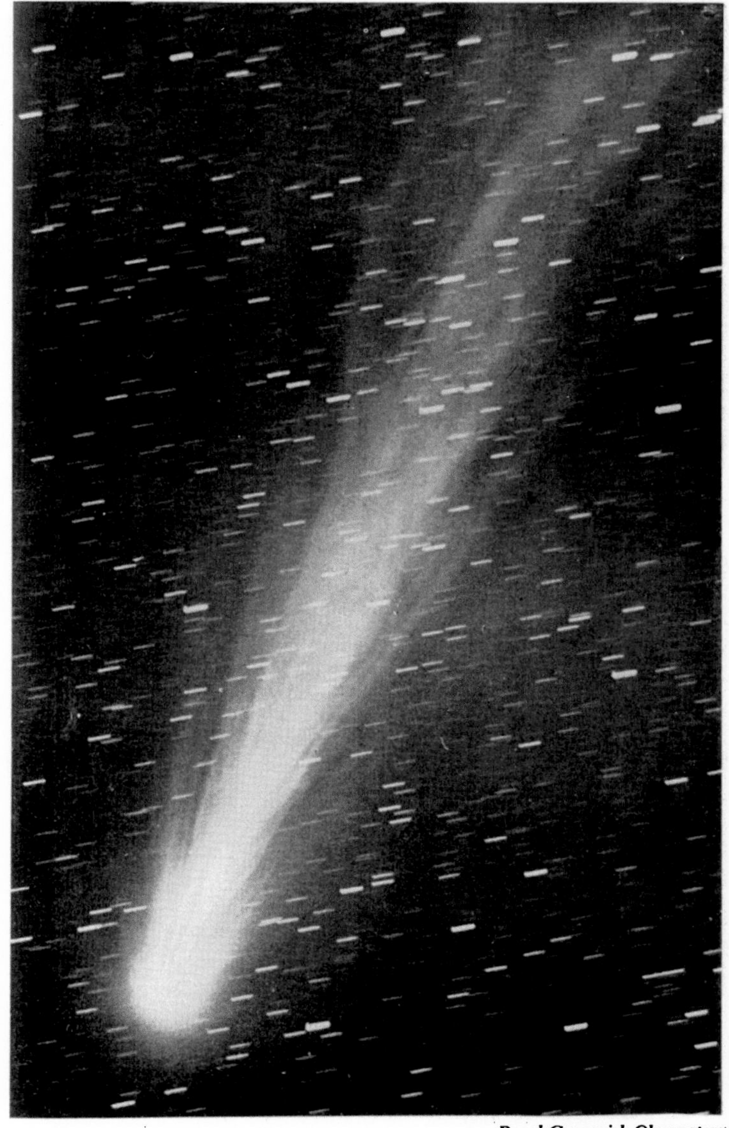

Royal Greenwich Observatory

Morehouse's Comet

PLATE IV

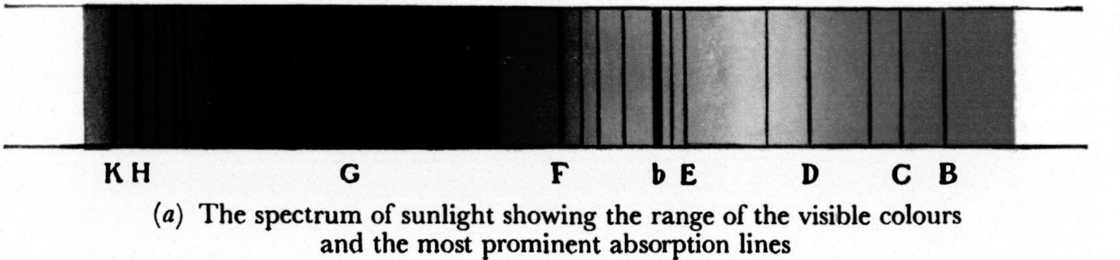

(*a*) The spectrum of sunlight showing the range of the visible colours and the most prominent absorption lines

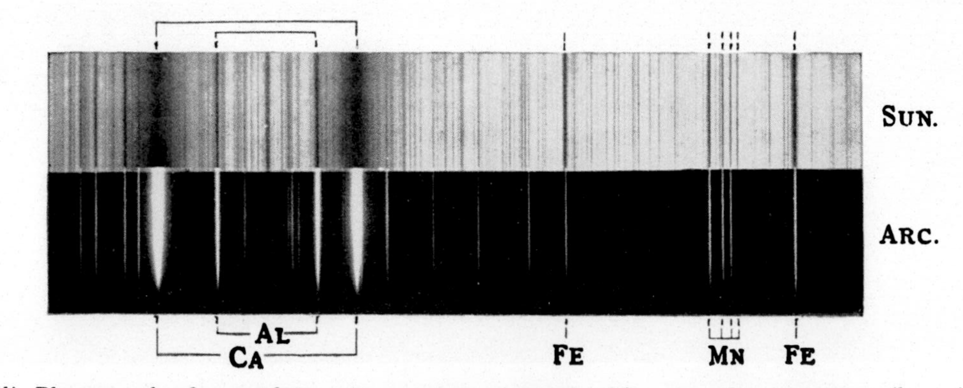

(*b*) Photograph of part of the solar spectrum (upper) with comparison spectrum (lower) produced by Aluminium (Al), Calcium (Ca), Iron (Fe) and Manganese (Mn)

sodium atoms from the heat of the Bunsen flame. It is found in a similar way that other elements have their own characteristic emission lines, varying in number and wave-length from element to element.

Experiment 2. If we examine the light from an electric bulb in the spectroscope, we observe the unbroken succession of the rainbow colours—this is called a *continuous spectrum.* Such a spectrum is characteristic of a luminous solid or of a glowing gas under great pressure.

Experiment 3. Now suppose that we place the Bunsen flame (in which a compound of sodium is suspended) between the electric bulb and the spectroscope so that the light from the bulb passes through the Bunsen flame now coloured a vivid yellow. In the spectroscope we shall now see the continuous spectrum (produced by the light of the bulb) and *two dark lines* (or *absorption lines*) with precisely the same wave-lengths as the emission lines in the first experiment. Evidently, the sodium atoms in the less intense flame of the Bunsen have the capacity to *absorb* the light-energy of the continuous spectrum in the two wave-lengths characteristic of the yellow light which can be emitted by the sodium atom.

The Sun's spectrum, as we have seen, consists of a continuous spectrum crossed by a multitude of absorption lines; two of these lines have wave-lengths 5890 and 5896 A.* Now the principles derived from our experiments suggest that the continuous rainbow spectrum is produced by the intensely hot radiating surface of the Sun—known as the *photosphere*—above which is a cooler atmosphere, although extremely hot by terrestrial standards, of which sodium is a constituent. When the solar spectrum is compared with the spectra of other elements examined in the laboratory, it is inferred that about sixty out of the ninety-two terrestrial elements are present in the Sun. Further, several refractory compounds such as magnesium hydride resist dissociation even at the very high temperatures

* These lines being close together are shown by a single line in Plate IV (*a*) above the letter *D*.

prevailing in the Sun's atmosphere or in the somewhat cooler conditions of sunspots; the absorption by the molecules of these substances gives rise to absorption bands in the spectrum.

The Influence of the Ozone Layer

The fact that about thirty of the terrestrial elements have not been identified so far in the Sun may suggest at first that these elements are actually missing in the Sun. But if an element is extremely rare in the Sun, the number of atoms in the solar atmosphere may quite well be so small that the absorption lines produced by them will be too faint to be recorded even by the most efficient equipment. Radium, for example, is a very rare element in the Earth and the absence of the absorption lines, peculiar to radium, from the solar spectrum makes it certain that, if radium is really present in the Sun's atmosphere, then it must be an extremely rare solar element. But one of the main reasons for the non-detection of most of the missing elements in the solar spectrum is the Earth's ozone layer. The molecules of ozone possess the property of absorbing radiations with wave-lengths less than 3000 A. approximately; thus most of the Sun's ultra-violet radiation is unable to penetrate our atmosphere. Now, the principal absorption lines of most of the missing elements lie in the ultra-violet range of wave-lengths and as this is a region that is hidden from us by the ozone layer we are unable to say definitely whether such elements exist in the Sun or not. The recent developments in jet-propulsion may make it possible in the future to send self-registering spectrographs into regions above the ozone layer and so enable us to learn the secrets of the solar ultra-violet radiation.*

The terrestrial ozone plays an important role in our daily lives. Ultra-violet radiation is, as is well known, used advantageously in medical treatment; but it must be used in small

* Since this was written experiments of the kind described have been made successfully.

doses, for otherwise its effect on the bodily tissues would prove disastrous. The human body, particularly the eye, has become adjusted through thousands of years to the range of solar radiations with wave-lengths between 3000 and 8000 A. but, through lack of opportunities for adaptation, is unable to withstand ultra-violet radiation in any great strength. The ozone layer is thus a buffer between the continuance of terrestrial life as we know it at present and sudden extinction by the short-wave rays of the Sun against which we have no physiological protection. We have no certainty of course that, over long spans of time, our ozone shield has remained constant in quantity and effectiveness; however, if the alteration in content is slow, it is reasonable to infer that life will be able to adapt itself successfully to the changing physical conditions.

We can now indicate briefly the explanation of the marked scarcity of helium in our atmosphere despite its continuous addition as a result of radioactive disintegrations. We have seen that atoms of sodium in a Bunsen flame can absorb radiations of two definite wave-lengths. Suppose that a particular atom absorbs the radiation corresponding to the wave-length 5890 A.; it is then charged with a certain amount of energy. It might be expected that the atom could continue to remain loaded with energy until it was forced to part with all or some of this energy, perhaps as the result of a collision with a neighbouring atom. But this is not the *normal* way of atoms: no sooner has an atom acquired this store of energy than it hastens spontaneously to get rid of it in the form of radiation appearing, spectrally, as an emission line; it is found, in fact, that the atom can normally remain energized for no more than about a hundred-millionth part of a second. This is true of other atoms and, in particular, of oxygen with which we are concerned for the explanation of the scarcity of helium in the atmosphere. But it is also possible for atoms such as oxygen to be energized in such a way that they appear very reluctant to disgorge their energy spontaneously; but when they do, the atoms being supposed not to have collided with others in the interval, the radiations are characteristic of the energy released. Such energized atoms are said to be in the metastable state.

It is known that the beautiful green and red colours of the aurora borealis (the 'northern lights'), or of the aurora australis, are produced by some of the oxygen, a hundred miles or more up in the atmosphere, in the metastable state; at these heights the atmosphere is so attenuated that such metastable oxygen has time to get rid of its energy spontaneously before collisions occur. The same radiations have been detected by Lord Rayleigh in the extremely faint light coming from all parts of the sky on a moonless night; these radiations, which are distinct from starlight, give evidence of permanent auroral conditions high up in the atmosphere.

But some metastable atoms are not so fortunate as to escape collisions with other atmospheric atoms and their bottled-up energy is released, not as radiation, but in imparting energy of motion to the particle in collision. It is found that the velocity which can be transmitted to a helium atom as the result of a collision is somewhat greater than the velocity of escape and so we have a continuous process by which the helium is gradually driven out of the atmosphere possibly at very much the same rate at which it is produced by radioactive disintegrations. As the velocity transmitted to hydrogen is greater than that in the case of helium, hydrogen also will be driven off by this process. The velocities transmitted to the other atmospheric gases are much less than the velocity of escape; accordingly, they are not likely to be dissipated in the way described.

The Earth's Crust

The rocks of which the Earth's crust is composed are chemical compounds of which oxygen is the principal constituent, with silicon, aluminium and iron in order of diminishing abundance; all told, these four elements account for about seven-eighths of the crust, the actual percentage composition being about 46½% for oxygen, 28% for silicon, 8% for aluminium and 5% for iron. Next follow calcium (3½%), sodium (2¾%), potassium (2½%) and magnesium (2%); the balance of about 1¾% is contributed by the other familiar elements such as carbon, sulphur, lead, tin, silver, gold, and so on. The rocks are divided into two main classifications, (i) igneous rocks, and (ii) sedimentary rocks. The igneous rocks such as granite and basalt are the solidified forms of the molten, or potentially molten, material below the solid crust which has been either gradually forced towards the surface as a result of deep-seated disturbances or violently ejected by volcanic eruptions. The sedimentary rocks, on the other hand, are products of the disintegration of igneous rocks through mechanical and chemical action, with water and carbon dioxide as the chief agents. For example, the interaction of atmospheric carbon dioxide with the calcium compounds of the igneous rocks produces calcium carbonate, that is, chalk and limestone in its best known geological forms. The feldspar in the igneous rocks is transformed as a result of chemical action into silica (quartz is one of its familiar forms), the carbonate of potassium (soluble in water and so carried off eventually to the oceans), and clay. The solid products are slowly transported by rivers and glaciers to the oceans on whose beds they accumulate and are compressed, forming eventually the sedimentary rocks, of which the commonest are chalk or limestone, sandstone and shale. A noteworthy instance of the process of sedimentation has recently been revealed by Sir Leonard Woolley while making archaeological excavations at Ur in Babylonia. At a considerable depth below ground-level he came upon a layer of blue clay, 8 feet thick, sure evidence of an unusual inundation far back in the past and no doubt corroborating the Biblical tradition of the great flood associated with Noah and his Ark.

The process of the transformation of igneous rocks into sedimentary rocks is extremely slow; as we shall see in a subsequent chapter, it affords a means of estimating, perhaps not with very great accuracy it must be confessed, the ages of the various sedimentary strata. But one point may be noted here. If the chalk downs of southern England have been formed in the way described, it follows that at some time, presumably in the remote past, the south of England was under the ocean and was afterwards raised by some great convulsion of nature several hundreds of feet above sea-level. Similarly, one might expect that lands which at one time supported life in all its forms are now submerged by the sea. Evidently, permanence is not a characteristic of the Earth's solid exterior. Further evidence to this effect is found in the great rugged mountain formations such as the Alps and Himalayas where rock disintegration and erosion have progressed but a short stage as measured by the time-scale of geology; these are, then, comparatively new formations produced by great upheavals or crustal adjustments occurring in the not too distant past. The Earth's crust is thus a scene of unceasing change; the change may be slow as in the gradual wearing away of the uplands and the transportation of the debris of solid rocks to ocean floors, or it may be on a cataclysmal scale as in the sudden buckling or fracture of the crust.

A further feature of the Earth may be conveniently mentioned at this point; this relates to the increase of temperature with depth below the surface. The evidence derived from the measurement of temperature in coal-pits, oil-wells and deep borings is conclusive. If we take the Carnarvon well in South Africa as an example, it is found that for the first half-mile below the surface, a section consisting of sedimentary rocks, the temperature increases on the average by 1° C. in 70 feet; in approximately the next half-mile below the surface, consisting of igneous rocks (granite, etc.) the temperature increases by 1° C. in 137 feet on the average. This increase of temperature by one degree in so many feet is called the temperature-gradient. The actual value of the temperature-gradient is governed by the nature of the rocks, but it can be influenced by subsidiary factors such as the presence of hot springs and radioactive material in the immediate neighbourhood of the boring. It is evident that at no great depth the temperature of boiling water (100° C.) is reached or even surpassed: thus, the temperature of boiling water is reached at a depth of 7200 feet in the well at Long Beach, California.* Its full depth is close to 9000 feet and at the bottom the temperature is round about 120° C. The temperature of molten lava, which comes from much greater depths than those we have just mentioned, may be as high as 1200° C.

The general picture of the Earth that we obtain from this rather brief discussion of temperature is that of a body extremely hot, presumably, at its centre, with the temperature diminishing towards the surface; since heat in a solid flows from a region of high temperature to one of lower temperature, a process called the conduction of heat, the central regions of the Earth must be parting with heat to the cooler regions nearer the surface. Further, since the surface heat is gradually being radiated away into interplanetary space, the Earth as a whole must be cooling; this last conclusion is subject to one important qualification, namely, that the production of heat by chemical action and radioactive processes together with the acquisition of heat from solar radiation cannot altogether compensate for the loss by conduction and dispersal into space.

The fact that molten lava is ejected from volcanoes suggests at first sight that below the solid crust there is an inner fluid core at a very high temperature. It is well known that the temperature at which a solid melts increases as the pressure, to which the solid is subjected, increases. For example, under ordinary conditions the melting-point of ice is 0° C., the pressure to which the ice is subjected being the atmospheric

* This well shares with the Bakersfield well, also in California, the distinction of being the deepest well in operation.

pressure of 14 pounds per square inch at the Earth's surface.* In the same way the pressure at a depth of 10 miles below the Earth's surface is the weight of the corresponding column of rock—and the air—of cross-section 1 square inch. The density of the rock forming the outer crust is about 2¾ times the density of water and since the average density of the Earth as a whole is known to be 5½ times that of water it is evident that the density of the terrestrial material must increase towards the centre where the density is estimated to be about twelve times that of water. Although the precise way in which the density increases downwards is not known accurately, some rough idea of the pressures within the Earth can be obtained. Thus at a depth of 1000 miles, the pressure is approximately equivalent to three-quarters of a million atmospheres, or about 4800 tons per square inch; at a depth of 2000 miles the pressure is about 11,000 tons per square inch and at the centre about 22,000 tons per square inch.

The effect of an increase in pressure on a solid is to compress the molecules still more closely, while the effect of an increase in temperature is to increase their mobility. Thus, pressure and temperature operate in opposite directions so far as molecular mobility is concerned; it follows that ice, for example, which passes from the solid to the liquid state at 0° C. under the ordinary atmospheric pressure—the melting-point as usually defined—can exist in the solid form at higher temperatures if the pressure to which it is subjected is increased; in other words, an increase in pressure leads to the raising of the melting-point. At depths well below the rocky crust of the Earth the enormous pressure existing there will have the effect of raising the melting-point of the terrestrial material beyond the actual temperature prevailing at the depths concerned; consequently, the material will have essentially the properties of a solid. If for some reason the pressure is con-

* This pressure can be conceived as the weight of a vertical column of air, with a cross-section of 1 square inch, extending to the upper bounds of the atmosphere.

siderably reduced, the melting-point temperature may fall below the prevailing temperature; under these circumstances the material will pass from the solid to the liquid state, and as such may be forced to the surface as a lava flow. The outflow of molten lava is, consequently, not evidence that the interior of the Earth is fluid although, owing to the high temperatures prevailing, we may describe the material as potentially fluid.

What are the circumstances in which there may be a localized diminution of pressure? It is to be remembered that the Earth is not a perfectly rigid body or a stable structure. The occurrence of earthquakes is sufficient evidence that rock-strata a few miles down suddenly slip or fracture under the weight of the overlying load which, in oceanic regions, is ever increasing owing to the continuous deposition of sediments. Considerable diminution of pressure as well as increase of pressure can thus be brought about in limited regions below the Earth's surface and, if the melting-point of the apparently solid material is reduced below the prevailing temperature, the substance takes the liquid form; if circumstances are favourable it is forced upwards through cracks or through the vent of a volcano (if one is close by) to appear as molten lava disfiguring the landscape and possibly spreading destruction to human activities in the vicinity.

In recent years the study of earthquakes has developed into an active branch of science, seismology. When an earthquake occurs, the sudden fracture of the crust or the large-scale breakdown of stability at depths of perhaps several hundred miles produces two main sets of waves which are recorded on self-registering seismographs, one passing through the rocky crust and the other through the central regions. When our knowledge as to the density of the crust and the average density of the Earth as a whole is correlated with the information supplied by earthquake waves, together with other considerations, it is inferred that the Earth's central core, of about 2000 miles in radius, consists mainly of iron or a nickel-iron alloy; that outside this core is a shell, roughly 1000 miles thick, consisting of a heterogeneous mixture of the elements and probably merging rather indefinitely with the iron core; that above this shell is another, again about 1000 miles thick, consisting of the material which gives rise to the igneous rocks; that above this is the comparatively thin crust, one or two score miles thick, consisting of the solid sedimentary and igneous rocks, beyond which is the envelope of atmosphere.

From this general description of the chief physical characteristics of the Earth we can draw an important conclusion relevant to our principal theme. The Earth is a body undergoing continuous physical change; in particular, as we have seen, it is cooling so that the further back we go in time the hotter the Earth must have been; at some stage it must, presumably, have been molten with a capacity of retaining only the heaviest gases and the heavier metallic vapours. Can we go still further back to a time when the Earth was so hot that terrestrial matter could exist only in the gaseous form? These may seem bold suggestions but scientific inquiry into the past history of the Earth is founded on the inevitable processes of evolution which may be said to be governed mainly, in the present connexion, by the simple fact that an isolated body such as a planet gradually becomes cooler through loss of heat into interplanetary space; this process may be partly arrested, as we have mentioned previously, by chemical actions and radioactive changes, but these sources of heat are not inexhaustible and are unable, in the long run, to influence the one-way evolutionary march of the Earth in the sense indicated.

The Earth's Tides

Another feature of the Earth which we shall briefly notice here is the phenomenon of the tides. As is well known the tides are due to the attraction of the Moon and, to a lesser extent, of the Sun on the waters of the seas and oceans. The general tendency of the lunar attraction (Fig. 9) is to cause a 'heaping' of the water around the point *A* of the ocean at which the Moon is vertically overhead, and also at the point *B* on the globe, diametrically opposite to *A*; under these circumstances we say that high-water occurs at *A* and *B*; at the same time high-water occurs, but not so pronounced, at places on the meridians of *A* and *B*. At high-water the depth of the ocean is greater than the average depth by a few feet, but both the time of occurrence and the height of the tide are modified, usually very considerably, close to coast-lines. As the Moon travels across the sky westwards the locality of high-water moves westward, keeping pace with the Moon, so to speak; the depth of water at *A* gradually diminishes until, about $6\frac{1}{5}$ hours after the time of high-water there the depth at *A* will be as much below the average as it was above at high-water; this state of the tide corresponds to low-water at *A*.

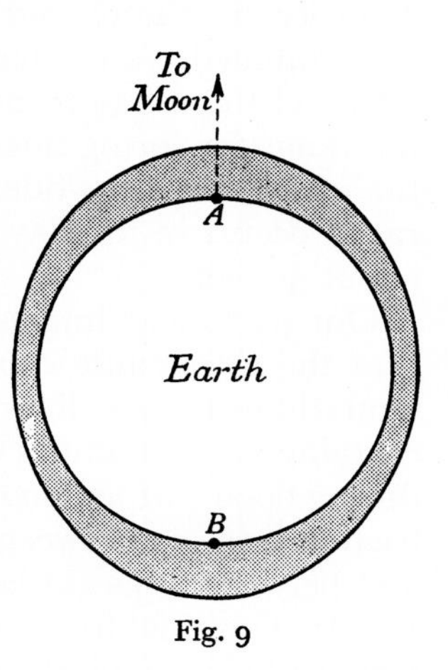

Fig. 9

The interval between high-water at a particular place and the next high-water is 12 hours 25 minutes on the average, and this is also the average interval between two consecutive low-waters. When the Sun's influence is considered as well, the height and the time of occurrence of, say, high-water depend on the relative positions of the Moon and the Sun. When these bodies produce high-water simultaneously the combined effect is called a spring-tide; this occurs at or near New Moon or Full Moon. When the relative positions of the Moon and Sun are such that the Moon is producing high-water and the Sun low-water at a particular place, the resultant tide there is called a neap-tide: this occurs about one week before and after New Moon. The tidal range at any point is the difference in the depths at high and low-water: at St Helena, for example—virtually corresponding to the open ocean—the average range is about 3 feet. But in estuaries and confined waters such as the Irish Sea and the Bristol Channel the range is much greater; in the Bristol Channel the range for spring tides is nearly 40 feet (at Cardiff) while the range for neap-tides is nearly 30 feet. The maximum range occurs in the Bay of Fundy where the spring-range is about 50 feet.

Our particular interest in tidal matters is due to the fact that the tides cause considerable movement of water, particularly near coast-lines where rapid currents are produced, carrying vast quantities of water into estuaries and channels during flood-tide and in the reverse direction during ebb-tide; friction is set up between the moving water and the sea-floor and between adjacent layers of water moving with different speeds. This tidal friction, as it is called, may not at first sight seem to be particularly important but, as we shall see later, it is one of those minute agencies which, operating over long intervals of time, produce significant results.

CHAPTER IV

GENERAL DESCRIPTION OF THE MOON AND PLANETS

THE principal surface features of the Moon—its mountains, craters, plains and so on—are well represented in Plate I, (facing p. 16). The most noticeable feature of the photograph is the considerable number of 'craters', large and small, the smaller ones in particular bearing a striking resemblance in form to the bomb-craters so tragically familiar in these recent years of 'total war'. The resemblance, however, stops there, for the lunar craters are on a far vaster scale than bomb-craters; the diameters of the latter are seldom more than a few score of feet, while the diameters of the former range from a mile or two up to 150 miles, the area of the largest crater being half as large again as the area of Ireland.

Another notable feature of the Moon's surface is the rugged character of the lunar scenery, with no suggestion of the weathering processes familiar in many of the Earth's hilly and mountainous areas. Now weathering, as we know it on the Earth, is a process primarily dependent on the continued existence of an atmosphere. Some of the gases of the terrestrial atmosphere, as we have seen, enter into chemical union with the exposed rocks; the never-ending cyclical process of the evaporation of the water of the oceans and elsewhere, and its subsequent condensation into rain provides a mechanical agency of erosion which, in conjunction with the chemical changes alluded to and acting over vast intervals of time, accounts mainly for the comparative smoothness of many of our hills and mountains.

Although the earlier cartographers of the Moon referred to many of the level expanses of the Moon's surface as 'seas',*

* In Plate I (facing p. 16) the great plain shown so clearly there is still referred to as the 'Mare Imbrium' (the sea of storms).

it is certain that the lunar surface is entirely solid, with not the slightest evidence of seas or of water in any aggregation, great or small. The absence of any indications of weathering on the lunar surface suggests that the Moon has possessed no atmosphere since the solidification of the crust and also that it is perhaps incapable of retaining one, in which event the absence of water is easily accounted for, since water, being easily vaporized, would soon escape into outer space. It may be said at once that both suggestions are accepted without reservation to-day; the evidence is based partly on observation and partly, as we have seen in Chapter III, on the principles of the kinetic theory of gases. The chief observational test is concerned with what is known as the occultation of a star by the Moon. As our satellite moves against the background of the stars it is continually passing in front of the stars lying in its path in the sky, thus temporarily blotting them from view—occulting them, as it is called. If we observe the occultation of a bright star—it would be difficult or even impossible to observe faint stars owing to the brilliance of the Moon—the disappearance of the star is instantaneous and not a gradual fading from view as it would be if a lunar atmosphere existed; on the latter supposition the rays from the star would be bent (or refracted) and scattered by the lunar atmosphere so that a diminishing amount of light from the star would continue to reach us for some time after the Moon had actually passed in front of the star, just as light from the Sun continues to reach us, as twilight, for some considerable time after our luminary has disappeared below the horizon. No such fading of an occulted star, which could be interpreted as a twilight effect, nor any gradual increase in brightness, on the reappearance of the star from behind the Moon, has ever been observed; the conclusion follows that a lunar atmosphere is non-existent or perhaps, more cautiously for the moment, that if the Moon has an atmosphere it must be extremely thin.

That the first alternative is correct is deduced when we

consider the molecular velocities of gases in a possible lunar atmosphere. From the known mass and dimensions of the Moon it is readily calculated that the velocity of escape from the lunar surface is about $1\frac{1}{2}$ miles per second. We have seen (p. 45) that if the average molecular velocity is one-quarter of the velocity of escape—say, $\frac{2}{5}$ mile per second for the Moon—the atmosphere will be dissipated in a few thousand years; also, we have seen that the average molecular velocity depends on temperature. Let us first consider temperature. The Moon is illuminated and heated by the Sun's rays which, at a given moment, fall on a particular half of the lunar surface. At full Moon we see the whole of a hemisphere and by means of a special instrument the temperature at the centre of the illuminated disk can be measured; it is about 120° C., that is, somewhat higher than the temperature of boiling water. This is a maximum temperature, for the temperature falls off at points at increasing distance from the centre of the bright disk; and at the centre of the dark disk turned towards us at New Moon the temperature drops to about −150° C. Reference to Table IV (p. 43) shows that the lightest gases, if they existed on the Moon, would be quickly dissipated at the moderate lunar temperatures now prevailing.

But, like the Earth, the Moon is slowly losing whatever heat it has and in the past it must have been hotter. According to a well-known theory, which we may adopt by way of illustration for the purpose in hand, the Moon was originally part of the Earth and later attained an independent existence when the latter was in a molten condition. If the corresponding temperature is taken to be of the order of 2000° C., the table of molecular velocities shows that none of the gases could be retained, for their average molecular velocities are all greater, some of them several times greater, than the critical velocity of $\frac{2}{5}$ mile per second. We conclude, then, that the Moon is a globe completely devoid of an atmosphere.

In these days of rockets and jet-propelled contrivances we

frequently hear of projects for visiting our nearest celestial neighbour and even for adventuring farther afield into planetary regions. When the recent extraordinary developments in scientific technology are borne in mind, such projects are perhaps not so fantastic as they at first sight may appear to be. Indeed there are, or have been in recent years, quite a number of 'Rocket Societies' founded for studying the possibility of extra-terrestrial travel; many of them have their own journals in which several of the relevant problems are discussed at length. Amongst these societies may be mentioned the British Interplanetary Society (its title suggests its ambitious aims); the Rocket Society of the American Academy of Science, founded as long ago as 1918; the Rocket Societies of Vienna, Moscow, Leningrad, Holland, and Berlin-Breslau, all founded between 1926 and 1938; several societies in South America; and that founded in Australia as recently as 1941. The German Society, it may be added, had over a thousand members and its own periodical publication. The Nazis appear to have recognized early the potentialities of the rocket as a weapon of war; the German societies had but a brief spell as scientific organizations, although the technical problems in which they were interested were taken over and successfully solved in secret by the German experts, the culmination being the invention of the V I and V II missiles which caused so much suffering and destruction during the final phases of the last war. A more recent development in this field of technology is the study of the possibility of utilizing nuclear energy for the propulsion of rockets, at the moment perhaps little more than a dream.

Several important requirements for an expedition to the Moon, for example, must not be overlooked. The absence of a lunar atmosphere would of course suggest to our prospective voyagers the necessity of taking an adequate supply of oxygen with them, just as our high-altitude airmen require to do. Again, we know that thousands, and perhaps millions, of meteors travelling at 20–40 miles per second bombard the

Earth every day; it is because of the protective barrier of the terrestrial atmosphere that these high-speed projectiles do not wipe out the human race or, at any rate, force us to live underground. Our lunar explorers will have no such protection and, if they hope to survive the incessant meteoric bombardment—for meteors must fall on the Moon in numbers comparable with terrestrial meteors—they must remember to arrange for a complete casing of armour-plate, at least several inches thick, resembling that worn by medieval knights in combat. But even this protection will be insufficient unless supplemented in another way: our explorers must also remember that on the Moon they will be exposed to the undiluted power of the Sun's ultra-violet radiation, for there is no ozone to shield them there; accordingly, they must arrange before setting out to have some means of protection from the perils inherent in sunlight. Perils of perhaps a more insidious nature arising from 'cosmic rays', the study of which is only now beginning to produce significant results, will certainly have to be guarded against. Altogether, the joys of space-travel are somewhat difficult to envisage.

The bombardment of the Moon by meteors throughout countless ages must have resulted in the lunar surface, at least its flatter parts, being carpeted with a layer of meteoric particles and dust, some of the latter being produced as a result of the fracture of the solid crust by the impact of the fast-moving meteors; this layer will remain undisturbed, for there are no winds on the Moon, except when a new meteoric arrival causes a temporary dust-storm round about the place of impact. The thickness of this layer of dust can only be a matter of speculation but, if we suppose that the meteors are, on the average, one-tenth of an inch in radius and that they have been falling on the Moon for a thousand million years at an average rate of three million per day, the average thickness of the dust layer over the entire surface of the Moon will be only about one-tenth of an inch; to this must be added some hypothetical amount resulting from the pulverizing of the crustal material; against all this must be set the loss of fragments bouncing outwards with velocities exceeding the velocity of escape. The outer skin of the lunar crust itself is probably volcanic material, for the almost complete incapacity of this substance to conduct heat is consistent with the observations of lunar temperatures made during the total eclipse of the Moon. Just before the partial phase begins the temperature of the central part of the illuminated disk is, as stated previously, about 120° C. and when the Moon has passed wholly within the Earth's shadow the temperature has dropped through more than 200°. It is evident that the effect of the Sun's heat can only be skin-deep and the inference is plain that the material forming the outer crust must be almost wholly incapable of conducting heat to even a moderate depth. Other observational evidence provides corroboration of the identification of the crustal substance with the familiar volcanic material on the Earth.

We can now proceed a stage further. Both the conclusion of the last section and the mountainous nature of the lunar surface point to violent changes in the crust similar to those experienced by the Earth's crust. We can look back into the distant past and see the Moon in a molten condition; gradually it cools and eventually a crust is formed. The lunar craters are supposed by some to be the evidence of immense volcanic activity during the last stages of solidification; but this speculation is prejudiced by the application of the word 'crater', borrowed from terrestrial associations, to the vast lunar formations unrivalled, so far as is known, on any other celestial body. Alternatively, it is suggested that the craters have been formed as the result of immense explosions following the penetration of the nearly solid crust by large meteors, perhaps as large as the smaller minor planets. Whatever the cause operating in the formation of the craters, we may be tolerably certain that the Moon must have been molten in the distant past. Further, the Moon's average density is $3\frac{1}{3}$ times that of water or only about 25%

denser than the crustal rocks of the Earth; it is probable then that the greater part of the Moon consists of rocky material similar to that forming the terrestrial crust.

Mercury

Mercury is the planet nearest the Sun, being about 36 million miles distant on the average; its orbit, however, is highly elliptical and its distance from the Sun varies actually between $28\frac{1}{2}$ and $43\frac{1}{2}$ million miles. The planet's surface, as viewed in the telescope, is generally featureless although some very faint markings have occasionally been observed—these are probably surface markings. A distinct marking would have been of great service in giving some indication of the planet's period of rotation; on other grounds it seems very probable that Mercury's rotational period is 88 days, which is also the planet's period of revolution about the Sun. If the orbit were circular, the equality of the rotation and revolution periods would ensure that one hemisphere would be turned perpetually towards the Sun, the other hemisphere being condemned to continuous darkness; however, owing to the considerable elongation of the orbit, only about three-eighths of the planet's surface fails to be illuminated at some time or other by the Sun's rays.

There is no observational evidence of an atmosphere. Nor should we really expect to find any substantial trace of one, for Mercury has but a weak capacity to hold even the heaviest gases at very high temperatures. The velocity of escape at the planet's surface is about $2\frac{2}{5}$ miles per second and, accordingly, if the average molecular velocity of a gas exceeds a quarter of this velocity, that is, about $\frac{3}{5}$ mile per second, the gas would almost immediately be dissipated; the critical velocity for the retention of a gas (one-fifth of the velocity of escape) is, for Mercury, about $\frac{1}{2}$ mile per second. It is found that the temperature of Mercury at points of its surface where the Sun is vertically overhead is about 400° C. on the average, being about 450° C. when Mercury is nearest to the Sun and about 365° C. when it is most remote. Reference to Table IV (p. 43) shows that, at the highest of these temperatures, hydrogen, helium, water vapour, methane and ammonia, and probably neon, would be quickly lost to the planet, that oxygen, argon, krypton, xenon and carbon dioxide, and probably nitrogen, would be securely retained.

But, like every other planetary body, Mercury is losing its original store of heat. The table shows that at 2000° C. the only gas, of those mentioned, that could be retained almost indefinitely is xenon, but at 5000° C. even this gas would be quickly dissipated. If, then, we can assign such high temperatures to Mercury in the past, it is almost certain that at the present time the planet lacks any substantial atmosphere, although the possibility of the formation of compounds such as carbon dioxide when the planet reached a comparatively cool condition, and their partial retention must not be lost sight of.

The planet's average density is about $3\frac{3}{4}$ times that of water, rather higher than that of the Moon. Remembering that the density of terrestrial rocks is about $2\frac{3}{4}$ times that of water, we are probably not very far wrong in inferring that Mercury consists mainly of rock and is similar in constitution to the Moon.

VENUS

Venus and the Earth are very much alike in size and average density. For Venus the velocity of escape is about $6\frac{1}{2}$ miles per second; consequently, the planet differs very little from the Earth in its capacity to retain an atmosphere. On the sunlit side of Venus the temperature can be as high as 60° C., while on the sunless side the temperature is as low as −20° C. That Venus possesses an atmosphere can be inferred in several ways. In the telescope the illuminated part of the planet is devoid of markings, a fact suggesting that we are prevented from seeing the solid surface by a blanket of clouds; this is further corroborated by the high reflecting power of Venus, for the solid surfaces of such airless bodies as the Moon and

Mercury are poor reflectors of sunlight. When Venus is in the crescent form a thin halo of light is seen to surround the unilluminated part of the disk, unmistakable evidence of the scattering of sunlight by an atmosphere, as in the familiar phenomenon of twilight; to this effect can be attributed the uncertainty of timing the beginning and end of the planet's transit across the Sun's disk, in the historic problem of determining the Sun's distance from the Earth by observations of such transits.*

Recently, spectroscopic observations have shown conclusively that Venus possesses an abundant atmosphere of carbon dioxide; both free oxygen and water have not been detected so far, and if they exist at all in the higher levels of the planet's atmosphere they must be present only in very minute quantities. On the other hand, it is not easy to escape the conviction that the clouds are very much like terrestrial clouds, composed of droplets of water in suspension at comparatively low heights—perhaps two or three miles—above the planet's surface.

The atmosphere and the clouds of Venus must act as an efficient blanket for retaining the considerable proportion of the solar heat falling on the planet; it is estimated that at the surface the temperature is about that of boiling water; under such conditions there is very little likelihood of the existence of expanses of water on the planet and it would seem that life, in any of the familiar terrestrial forms, must be wellnigh impossible at present and for a very long time to come.

MARS

In many ways Mars is the most interesting of the planets, for it is the only one whose surface can be observed and studied in detail. The planet's diameter is rather more than half the Earth's diameter; its average density is just about four times that of water and about three-fourths that of the Earth. When

* Transits of Venus are infrequent phenomena; the last two were in 1874 and 1882; the next two will occur in 2004 and 2012.

the Sun is vertically above a point on the Martian equator—that is, at the Martian noon—the temperature there is about 20° C., approximately the spring temperature at noon in England. Away from the equator the noonday temperature diminishes until in the polar areas white expanses are encountered, the so-called polar caps, suggestive of the Arctic and Antarctic ice and snow-fields of the Earth. At the Martian midnight the temperature is a score or two of degrees below zero, even in equatorial regions.

The velocity of escape from Mars is 3·1 miles per second; from Table IV (p. 43) it is easily inferred that only hydrogen and helium could not be retained at the temperatures now prevailing. Despite the possibility of very high temperatures in the past we might expect the planet to have retained some small amount of atmospheric gases at least; this expectation is not disappointed, for Mars has a transparent atmosphere, the existence of which was first inferred from occasional observations of what appeared to be clouds; also the shrinking and subsequent growth of the polar caps during the progression of the Martian seasons is suggestive of atmospheric transfer of water-vapour or carbon dioxide—if the caps consist of ice or frozen carbon dioxide, their most likely constituents. In 1924 photographic observations of the planet made at the Lick Observatory furnished complete corroboration of the existence of a Martian atmosphere.

The principle of the method employed depends on the relative penetrating powers of light of different wave-lengths and can be illustrated by means of familiar terrestrial phenomena: (i) the blueness of the sky when the Sun is high, and (ii) the orange or reddish hue of the Sun at sunset. The two phenomena are explained in terms of the property of the atmosphere to scatter light; it is known that violet and blue light is scattered to a very much greater extent than orange and red light and also that the degree of scattering depends on the thickness of the atmosphere through which the light passes; further the penetrative power of blue light is less than that of red light. In Fig. 10 OA is the vertical height of the atmosphere for an observer at O, and OB is the direction of the horizon corresponding to sunset.

Consider the first phenomenon and suppose, for simplicity, that the Sun is vertically above O in the direction of OA. Light of all wave-lengths reaches the observer at O—some is scattered of course in its passage from A to O—and the Sun appears white. But sunlight is also falling all over the hemi-

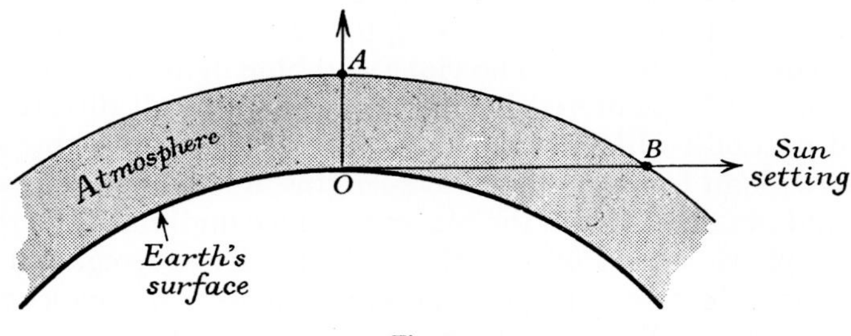

Fig. 10

spherical shell of atmosphere facing the Sun; the violet and blue components are scattered in all directions, the other colours to a minor extent, and some of this light will reach us from all directions in the sky, giving the impression that the sky is blue.

Consider now the second phenomenon. The thickness of the air through which the sunlight passes to the observer at sunset is represented by OB, and this is a much longer path than OA. Consequently, a much greater proportion of violet and blue light will be scattered between B and O than between A and O, and so the light from the setting Sun will be substantially deficient in the violet and blue wave-lengths. On the other hand the passage of the orange and red light is not nearly so seriously affected. Thus at sunset the sunlight is robbed of most of its violet and blue light, and of the adjacent colours in rapidly diminishing proportions, the balance of white light is upset and the Sun is seen mainly in orange and red light giving normally a golden appearance. The effect is markedly enhanced when the atmosphere is charged with water vapour or, in the desert, when sand-storms fill the air; only the longest wave-lengths succeed then in passing through and the Sun appears a vivid red.

Let us now consider the application of these principles to Mars. Since we can see the surface markings, the red component of sunlight, at least, must be able to penetrate the Martian atmosphere, if it exists, to the planet's surface where it is reflected, retraversing the atmosphere and finally entering our telescope. The violet and blue light on the other hand, will be scattered by the atmosphere in all directions and part of it will enter our telescope. If we could segregate the red light from all other wave-lengths, it is evident that we should obtain a view of the planet's surface undimmed by the atmospheric effects of the other colours. This segregation is simply achieved by placing a sheet of red glass in the telescope, for red glass will permit the passage of red light only. If this red glass is placed in front of a photographic plate we obtain a photograph of the planet's surface; such a photograph we can briefly describe as a 'red photograph'. In the same way, we can place a sheet of violet glass in front of another photographic plate and the record on the plate will be made by violet light only which alone can pass through the plate; we refer, in short, to this photograph as a 'violet photograph'. As the violet light comes to us from all parts of the upper levels of the planet's atmosphere, the 'violet photograph' is essentially a photograph of the outer atmospherical surface.

In Plate V*b* (facing p. 76), a 'red' and a 'violet' photograph of Mars are shown. The former shows only the most prominent surface markings because for various reasons we cannot expect faint markings to be recorded on a photograph of such a small object as Mars is in the telescope. The latter is entirely featureless except for the polar caps which now are revealed, partly at least, as atmospherical manifestations; these latter are probably thin clouds high above the polar

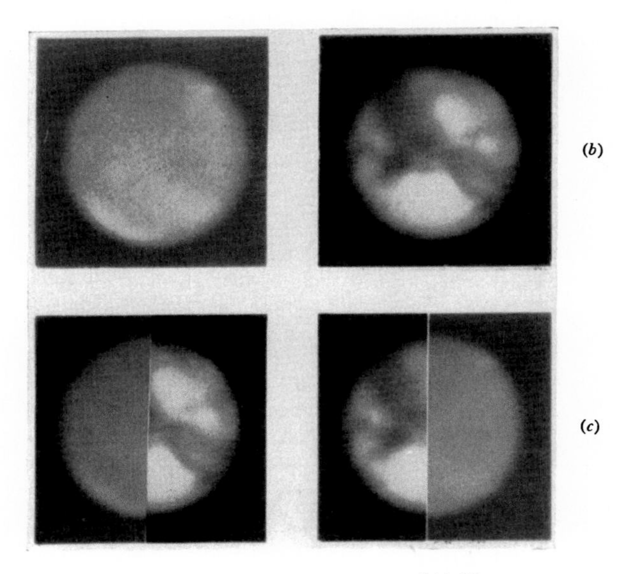

Lick Observatory

(*a*) Mars in ordinary light.
(*b*) Left: Mars in violet light.
Right: Mars in red light.
(*c*) Halved images of (*b*) for comparison

regions, although it is difficult to understand why such clouds should remain anchored, as it were, above the poles and nowhere else. The 'red' and 'violet' photographs provide one significant result: the size of the former, being a record of the surface, will indicate the radius of the solid globe; the 'violet' photograph, being a record of the high atmospheric levels, will be larger than the 'red' photograph, the difference in radii of the two photographs being ascribed to the height of the atmosphere. The difference in radii is shown very clearly in Plate V*c*, where halves of the 'red' and 'violet' photographs are placed together. It is easily inferred by measurement of the photographs that the height of the Martian atmosphere is about 60 miles. Further, it can be calculated that the density of the atmosphere at the surface of the planet is approximately that of the terrestrial atmosphere at the summit of Mt Everest, that is, at a height of about 5½ miles; at this height the air is so tenuous as to be incapable of supporting human life.

The composition of the Martian atmosphere is not fully known. Spectroscopic investigations have shown that water-vapour is present, but not extensively; oxygen has not been detected so far, although there are grounds for believing that it must have been a constituent in the past. Recently Dr G. P. Kuiper has detected carbon dioxide to an extent somewhat similar to that found in the Earth's atmosphere. It may be added that he has failed to discover any trace of ammonia gas and methane, two gases which, as we shall see later, are constituents of the atmospheres of Jupiter and Saturn.

The principal large-scale features of the Martian surface are the great coloured expanses, some grayish blue and the remainder reddish orange. The latter are almost certainly similar to our great terrestrial deserts. The former undergo seasonal changes in colour, suggesting the cyclical growth and decay of vegetation so familiar on the Earth. The discovery in 1877 of the so-called 'canals'—more accurately, 'channels'—by the Italian astronomer Schiaparelli stimulated a more lively interest in Mars and also a considerable amount of speculation on the possibility of the existence of intelligent beings on the planet. The 'canals' were described as faint tracks, several scores of miles, at least, in width and pursuing undeviating courses for hundreds of miles over the planet's surface irrespective of presumed changes of level, for we cannot but suppose that the Martian scenery is not without its mountains and valleys. They were supposed by Lowell to have been constructed for purposes of irrigation by intelligent beings, still continuing to survive on a globe sadly deficient in adequate supplies of water and far spent in its evolutionary life.

Although a few of the most prominent 'canals' have been photographed on favourable occasions, it is now generally believed that the fainter canals, 'discovered' by Schiaparelli and later observers of Mars, have no objective existence. It is to be remembered that, even in the largest telescope, Mars is but a small object and there is naturally a limit to the capacity of the eye in seeing small and faint markings. Many experienced observers endowed with the keenest vision, as indubitably substantiated by their work in other fields of observational astronomy, have failed to see the 'canals' in dispute. It has been stated that a considerable part of the observing time with the great 200-inch telescope on Mt Palomar in California will be allotted to the study of the planets and it may be confidently expected that our knowledge of Mars in all its aspects will be very considerably increased. Meanwhile we take our leave of Mars, to the best of our knowledge a dying planet, far more advanced in physical evolution than the Earth, and still more so than Venus, and approaching by slow stages the lifeless state of the Moon.

JUPITER

The appearance in the telescope of Jupiter, the largest and most massive of the planets, is shown by the photograph in Plate II*b* (facing p. 17); the photograph, however, does not give any indication of the delicate and variegated tints of the planet's exterior. Two features immediately present themselves to the eye: first, the unmistakable flattening, evidence of the rapid rotation of the planet; second, the system of markings in belts which are roughly parallel to the planet's equator, as may be inferred from the motion of any marking resulting from rotation. The periods of rotation of different markings vary from $9^h\ 50^m$, at or near the planet's equator, to about $9^h\ 56^m$ in high latitudes; consequently, what we see cannot be the solid surface of the planet, but the outer regions of an atmosphere. Most of the markings last but a comparatively short time—from a few days to a few weeks or occasionally to a few months; several, however, notably the great Red Spot, as it is called, exist for a considerable number of years. The short-lived markings are undoubtedly localized clouds but not, as we shall see, of water-vapour; on the other hand it is difficult to account for the long-lived markings which must, presumably, be related to some unknown feature of the planet of a semi-permanent character.

The velocity of escape for Jupiter is 38 miles per second. At temperatures considerably higher than 5000° C., even hydrogen could be easily retained; so even without the observational evidence of an atmosphere provided by the varied rotational periods of the markings, we would be justified in believing that the planet is enveloped in an atmosphere, probably of great density. We have already alluded to the preponderating abundance of hydrogen in the Universe and, in the circumstances pertaining to Jupiter, we might reasonably expect to find the Jovian atmosphere rich in hydrogen, if not in the free state then in chemical compounds. In recent years spectroscopic observations have shown conclusively that, to the depths explored, the atmosphere contains ammonia and methane, both most unpleasant gases, the first a compound of hydrogen and nitrogen, and the second a compound of hydrogen and carbon; methane, it may be added, is sometimes known as marsh-gas and sometimes as fire-damp, the deadly and explosive gas so feared in coal mines.

The temperature of Jupiter is extremely low, about $-120°$ C. and, evidently, if water exists on Jupiter it must be as ice; since oxygen is abundant in the Universe, it is perhaps reasonable to suppose that ice is an important constituent of the planet. At comparatively small depths below the gaseous exterior of Jupiter the considerable pressure together with the low temperature would ensure that all gases would be liquefied or solidified; it would then seem that, if even the outer shell of the planet consisted of elements or compounds which are classified as gases under normal terrestrial conditions, it is only the outermost fringe that is really gaseous, consisting mainly of the two gases ammonia and methane, the former being close to its point of liquefaction at the low temperature prevailing. The markings are then, probably, clouds formed of droplets of liquefied ammonia in suspension at the highest levels.

As we have seen, the average density of the Moon and the planets so far described range from about $3\frac{1}{3}$ times (for the Moon) to about $5\frac{1}{2}$ times (for the Earth) that of water. The average density of Jupiter is only $1\frac{1}{3}$ times that of water, thus presenting us with a problem of constitution very different from that relating to the Moon and the inner planets. It is at once evident that, unlike the bodies mentioned, the main bulk of Jupiter cannot consist of rock, of which the density is about $2\frac{3}{4}$ times that of water, or of heavier substances. Without going into further details we give what is believed to be, broadly, the constitution of Jupiter: the central core with a diameter of about 40,000 miles—about half the diameter of the planet—consists principally of rock (if metals are present, they must be in comparatively small proportions); outside this core is a shell of ice about 16,000 miles thick and above this the atmosphere, of small density, to a height of about 6000 miles. It is to be remembered that atmosphere in this sense implies, except at the extreme upper levels, the existence, in the liquid, or even the solid state, of those elements and compounds which are normally gaseous under ordinary terrestrial conditions. Further, it is evident that if a body is built up according to the specifications mentioned so as to have the known average density of Jupiter, the atmosphere must consist of substances with densities very much less than the density of water; only hydrogen, helium, ammonia and methane in the liquid or solid state fulfil these conditions and it is, accordingly, inferred that the atmosphere is constituted mainly of these elements and compounds.

Nothing is known about the temperatures prevailing within the rocky core of the planet. The interior of the Earth is, as we have seen, intensely hot and also, owing to the poor conducting properties of its material, it is losing its heat very slowly. These considerations are likely to apply with equal or greater force to Jupiter, for its central core is no less than five times larger in diameter than the Earth. Perhaps the spasmodic release of its bottled-up heat is responsible for the more distinctive changes apparent in the upper levels of the planet's atmosphere.

Of the four great satellites of Jupiter, the only ones we need consider in the present discussion, two (Io and Europa) are rather similar to the Moon as regards diameter and mass, and the other two (Ganymede and Callisto) are rather larger than Mercury but less massive; the velocities of escape for these bodies are intermediate between the velocities of escape for the Moon and for Mercury, nearer the former on the whole. As we have found no evidence for an atmosphere on the Moon and on Mercury, we might at first conclude that these satellites are also devoid of atmospheres. But the temperatures of the satellites must be extremely low, lower even than the temperature of Jupiter; it is possible that these bodies have retained a modicum of atmosphere presumably of the heavier gases, although there has been no direct observational evidence so far in support. However, as we shall see, an atmosphere has been detected on Titan (the largest of

Saturn's satellites), and this body and the four great Jovian satellites have points of general similarity as regards size and mass.

Saturn

Saturn, unique amongst the planets for its marvellous system of rings, is justly regarded as one of the most beautiful objects in the heavens (Plate II *c*, facing p. 17). We first consider the globe of the planet. Somewhat smaller than Jupiter, it is very much flattened, even more so than Jupiter, and we ascribe the flattening to a rapid rotation about the smaller diameter. Like Jupiter, Saturn has a system of bands but so indistinct and featureless that they are of little help in studying the rotation of the globe; however, 'spots' have been observed very occasionally, thus enabling the period of rotation, which is $10\frac{1}{4}$ hours, to be determined. The velocity of escape from Saturn is nearly 23 miles per second; accordingly, the continued retention of an atmosphere is not in question, just as in the case of Jupiter.

Spectroscopic observations reveal the presence of ammonia and methane, the former not in any great abundance. As the temperature of Saturn's globe is about $-140°$ C., the comparative rarity of ammonia is largely due to the fact that most of the gaseous ammonia is liquefied or solidified at the temperature mentioned; the absence of distinct markings on Saturn, as compared with Jupiter, can probably be ascribed to this difference in atmospherical content.

Perhaps the most remarkable feature of Saturn's globe is the smallness of its average density, which is only seven-tenths that of water and notably smaller than the density of any other planet. It would therefore seem that a considerably larger proportion of the volume of Saturn than in the case of Jupiter must be ascribed to an 'atmosphere', or outer shell, of liquid or solid hydrogen, helium, ammonia and methane all of which, as we have already remarked, have densities in these states very much less than the density of water. The constitution of Saturn's globe is believed to be somewhat as follows: a rocky core, 28,000 miles in diameter, above which is a shell of ice about 6000 miles in thickness, above which again is the 'atmosphere' about 16,000 miles in height with an average density one-quarter that of water and consisting mainly of liquid or solid hydrogen and helium and to a lesser extent of ammonia and methane. Allowing for variations in constitution the build-up of Saturn resembles that of Jupiter in its large-scale features, and the further general inferences made with respect to Jupiter are also applicable to Saturn.

Of Saturn's nine satellites Titan is the only one of considerable size which need be considered as regards the possibility of the existence of an atmosphere. Its diameter is about 18% larger than that of Mercury but its mass is only a little over half that of this planet. The average density of Titan is about $1\frac{1}{5}$ times that of water—just a little less than the average density of Jupiter. The velocity of escape for Titan is $1\frac{3}{5}$ miles per second, a little greater than the velocity of escape from the Moon and not quite three-quarters that for Mercury. However, the temperature of Titan must be extremely low and the present retention of the heavier gases is possible. Actually, methane has recently been detected and the presence of ammonia is suspected; possibly, the atmosphere is very tenuous and, presumably, it was formed after Titan had reached a sufficiently cool state during its evolutionary progress. The low density of Titan, taken in conjunction with a thin atmosphere, suggests that Titan has a rocky core with a diameter about half that of the whole globe, the remainder of the globe being ice.

We now consider briefly the constitution of Saturn's rings. Fig. 11 shows a part of the ring-system as we would see it looking at right angles to the plane of the rings. However constituted, the rings must be very thin for they are invisible on those occasions when the Earth lies in their plane. It was proved by Clerk-Maxwell in 1856 that, under the gravitational forces to which they are subjected, the rings could neither be liquid nor like solid sheets for, on either hypothesis, the stability of the rings was impossible. He went on to show that the rings must consist of vast swarms of small satellites, each revolving in an approximately circular orbit according to the law of gravitation. Just as in the planetary system the orbital speeds of the planets diminish in a precise manner from Mercury outwards, so the orbital speeds of the ring-satellites

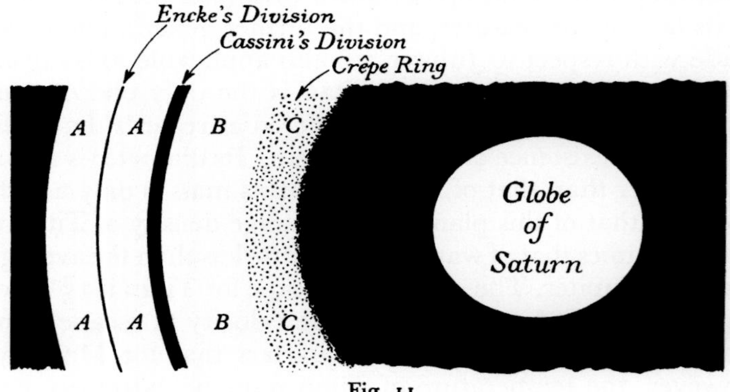

Fig. 11

must diminish in a similar way from the inner boundary of the ring-system towards the outer boundary; this was later verified by spectroscopic observations.

As Fig. 11 shows the rings are not continuous; the principal bright rings, which are denoted by *A* and *B*, are separated by a gap known as Cassini's division (first detected by Cassini in 1675) which is about 3500 miles in breadth. Later, it was discovered by Encke that ring *A* is not continuous; the narrow and faint division shown in the figure is known as Encke's division. Ring *C*, known as the crêpe ring from its dusky appearance, was discovered in 1850; it is a faint extension from the inner edge of the bright ring *B*. Cassini's division is interpreted as a region in which ring-satellites are not found; this is corroborated by observation and by theory. Suppose that a satellite is pursuing an orbit within the Cassini division; it would, of course, be attracted by the globe of the planet, by the ring-system and also by the planet's satellites particularly by Mimas, the satellite nearest the globe of the planet. The result of calculations shows that such an orbit would not be stable, for the ring-satellite would be constrained to move either farther away from, or nearer to, the globe; Cassini's division is thus a zone of clearance. Encke's division is explained in a similar way. The crêpe ring is evidently a zone in which the number of ring-satellites is comparatively small.

What is the origin of Saturn's incomparable system of rings? If we suppose that some time in the past a large satellite moved in an orbit very close to the globe of the planet then, as a result of the gravitational attraction of the globe, the parts of the satellite nearer the globe would be subjected to much greater strains than parts farther away; the satellite must then be fractured and in the course of time broken up into thousands of small portions which eventually spread out into the ring-system as we see it to-day. Associated with any planet, then, there is a danger-zone within which a satellite runs the risk of being broken up; the extent of this zone, in relation to the diameter of the planet, is known as *Roche's limit*.

Uranus, Neptune and Pluto

As we have not a great deal of information about these planets it is convenient to discuss them together. The velocities of escape for the first two are $13\frac{1}{2}$ and $14\frac{1}{5}$ miles per second respectively; accordingly, Uranus and Neptune are easily capable of retaining an atmosphere. The temperature of Uranus is $-180°$ C. approximately and that of Neptune is about 30° C. colder. Spectroscopic observations show that these two planets have an atmosphere of methane; there is no trace of ammonia although, if it exists in the planets, it is certain to be in the solid state, frozen out of the atmosphere owing to the extreme cold prevailing. As the average densities of Uranus and Neptune are almost identical with that of Jupiter, it is probable that they resemble Jupiter closely in physical constitution.

Very little is known about Pluto; it is certainly a small planet, perhaps about the size of Mercury or Titan and perhaps, like Titan, it has some trace of an atmosphere of a similar constitution.

Meteorites

As we have seen in the previous chapter, meteorites are solid chunks of cosmic matter arriving on the Earth's surface from interplanetary or interstellar space. Their chemical composition and physical characteristics are readily ascertained by ordinary laboratory methods. It is found that practically all meteorites contain a nickel-iron alloy in varying proportions and their classification is based on the relative amounts of this alloy. Meteorites which consist almost entirely of nickel-iron are known as *siderites* or, simply, as iron meteorites; the analysis of over 300 of these objects shows that the percentage of nickel-iron in their composition is a little over 99%. Meteorites mainly rocky in composition and with comparatively little nickel-iron are called *aerolites*, or stony meteorites. Those in which rock* and nickel-iron are in more or less equal proportions are known as *siderolites*, or stone-iron meteorites. There is a fourth class called *tektites*; these are small glassy objects; little is known definitely about their origin.

In the past, many speculations as to the origin of meteorites have been offered. Laplace considered the possibility that they had been ejected in the distant past from lunar volcanoes, while other astronomers surmised that they had been ejected from terrestrial volcanoes; some associated meteorites with comets, some with the debris of the supposed disruption of the Moon from the Earth. These speculations, of course, rested on very little scientific evidence. In recent years a certain amount of definite information has been gathered from which it is possible to draw a little nearer to the solution of the problem.

* Mainly the silicates of the commoner metals together with small proportions of non-metallic elements such as carbon, phosphorus and sulphur.

Whatever their classification, all meteorites reveal the common characteristic of a previous igneous state; in other words, their material was unmistakably in a molten condition at some time in the past. One significant feature of the iron meteorites is their peculiar crystalline structure, not encountered elsewhere in nature. Recent laboratory experiments suggest that this structure is almost certainly the result of very gradual cooling from the molten state under extremely great pressure. In the cosmic sense meteorites are very small bodies and the pressure within the largest is too trifling to produce the crystalline structure observed. It would then appear to be certain that such meteorites are fragments of very much larger bodies, probably as large as the biggest of the minor planets at least, which have suffered disruption in the distant past. In a later chapter (p. 154) we shall explain how the age of a meteorite can be accurately determined, the age referring to the interval since solidification occurred; meanwhile, it is sufficient to record here that the ages vary from about 100 million years to nearly 3000 million years. Not all iron meteorites show the characteristic crystalline structure mentioned: either such meteorites never possessed this feature, in which event it is possible that they represent fragments of the outermost parts of a larger body of planetary dimensions; or, if they did possess it, a subsequent melting of these meteorites, such as would occur if they passed comparatively close to the Sun at some moment in their cosmic existence, removed all trace of the crystalline structure.

From what has been said, particularly with reference to the suggestion that meteorites are fragments of bodies of planetary dimensions, it may seem to be inferred that these bodies have originated in the Solar System; although the balance of opinion is in favour of this view, the possibility of the origin of some meteorites in extra-planetary space cannot be ruled out. As we have seen, much can be learned about meteors from the observations of their luminous tracks in the sky. Although the distinction between meteors and meteorites is mainly confined to a difference in size, no observations of the fall of meteorites* have so far yielded precise information as to their velocity before hitting the Earth. If the velocity proved to exceed 26 miles per second, there would be substantial grounds, although not entirely conclusive, for asserting that the meteorites entered the Solar System from extra-planetary space; if the velocity proved to be less than that just stated, there would be equally substantial grounds for regarding meteorites as members of the Solar System, as is the case for the majority of meteors. We take leave of meteorites for the present, with no certainty as to their place of origin but with some assurance as to the physical circumstances attending their birth.

* It is reckoned that, on the average, five are seen to fall annually.

CHAPTER V

A PROVISIONAL SUMMING-UP

WE have now arrived at a position when we can usefully attempt to supply an answer to the first of our main questions as to the origin of the Earth and the planets. In the survey of the preceding chapters, particularly Chapter II, we have suggestive evidence that the Solar System is an organic unity and that the planets are sprung from a common ancestor; the only alternative that bears scrutiny insists that the planets resemble adopted children picked up individually, as it were, by the Sun at different times and, possibly, under very diverse circumstances. The overwhelming argument against the second alternative involving, as it does, the piecemeal capture of the planets by the Sun is found in the uniformities discernible in the orbital and rotational motions of the principal members of the Solar System. First, the paths of the planets around the Sun lie generally in planes very little different from the plane of the ecliptic (the plane of the Earth's path around the Sun) and all the planets, major and minor alike, revolve in the *same* direction in their orbits.

Further, the directions in which the Sun and planets rotate about their axes are, in general, the same. It must not, however, be supposed that the equatorial planes of the Sun and the planets bear any simple relation to the plane of the ecliptic: the Sun's equator, for example, is inclined at an angle of about 7° to the ecliptic; the corresponding angle for the Earth is $23\frac{1}{2}$°, and so on; the case of Uranus is remarkable, for its equator is nearly perpendicular to the ecliptic and the direction of its rotation is exceptional being, in the strict sense, opposite to the direction of rotation of the Sun and the remaining planets where this direction is known. It is to be remarked, in this connexion, that no information has so far been obtained as to the direction of rotation of Pluto or of the swarms of minor planets. It must be confessed that it is difficult to account for the exceptional circumstances relating to Uranus if we regard, as indeed we do, the uniformities of orbital and rotational motion in general as providing an incontrovertible argument in favour of the common origin of the planetary system; it is straining the probabilities too far to suppose that the capture of individual planets by the Sun occurred with such fine discrimination that only those revolving and rotating in a common direction became permanently associated with the Sun.

When we turn to the satellite systems we have evidence of a similar nature, up to a point, although not nearly so consistent as in the case of the planetary motions. We know the direction of rotation of the Moon about its axis—it is the same as that of the Sun and planets—but we have little information about the sense of rotational direction for the other satellites. We are thus reduced to considering only the evidence furnished by the orbital motions of the satellites around their respective planets. As we have seen, the great satellite systems of Jupiter and Saturn (see Figs. 6 and 7, p. 30) are, to some extent at least, miniature replicas of the planetary system itself. But the resemblance breaks down in one important particular, for the three outermost satellites of Jupiter and the outermost satellite of Saturn are exceptional as regards the direction of orbital motion around their respective planets; we recollect that the orbital motion of these satellites is retrograde, that is, opposite to the direction of motion of all the other satellites in these systems, the latter circulating around Jupiter and Saturn in the same direction as that of the planets around the Sun. The retrograde satellites thus spoil our picture of uniformity. But the significant feature is that these satellites are at very great distances from their respective planets as compared with the distances of the inner or normal satellites; they are, accordingly, much more loosely held in control by the gravitational attraction of the planets than are the inner satellites. This fact suggests that Jupiter, for example, has acquired the retrograde satellites by a process essentially different from that appertaining to the normal satellites. This process may be the capture of these small bodies, possibly minor planets originally, by the massive Jupiter and the normal members of its satellite system. It is necessary then to consider the problem of 'capture' in some detail.

It must be remembered that when we speak of the orbit of a planet around the Sun, or of a satellite around its planet, as being an ellipse, we are implicitly assuming that only the Sun's attraction on the planet—or the planet's attraction on the satellite—is operative and we are, in fact, disregarding the attractions of all the remaining bodies of the Solar System. But in determining the accurate path of a planet such as Mars around the Sun we can regard the elliptic orbit as only a close approximation to the real path, the resulting modifications being due to the attractions of the other planets; it can readily be appreciated that the dynamical theory of planetary motions is a complicated and difficult subject. Normally, in the planetary system, the differences between the actual path of Mars, say, and an ellipse are not significantly large, simply because, first, the bodies concerned are all well spaced out and, second, the planetary masses are very small compared with the Sun's mass, with the result that the Sun exercises a predominant gravitational control. But this is no longer true if a body such as a comet can approach close to a planet such as Jupiter for, since the attraction of one body on another depends on their distance apart, increasing as the separation diminishes according to the inverse-square law, it is possible for the attraction of Jupiter on the comet to exceed temporarily the attraction of the Sun on the comet. The result of such a close encounter would undoubtedly be a very substantial change in the character of the comet's orbit; in the case of Morehouse's Comet, as we have seen, the effect of the attraction of one or more of the major planets has been to break the bond of solar control, the comet being now lost to the Solar

System. The reverse process is also possible; if a comet, entering the Solar System from interstellar space, passes close to Jupiter, for example, its path may be transformed into an elliptic orbit around the Sun and the comet then becomes a member of the Solar System; this is an example of the process of 'capture'.

A second example of the possibilities of capture is furnished by the minor planets circulating round the Sun between the orbits of Mars and Jupiter. Many of the orbits of these bodies are considerably elongated and it may well have happened at some time in the past that a minor planet found itself in the close proximity of, say, Ganymede, the most massive of Jupiter's satellites; it is certain that under these circumstances its orbit would be considerably altered and, if the conditions were favourable, it might be captured by the planet, thereby having its status changed from a minor planet to a satellite of Jupiter; in this event its subsequent orbital motion around Jupiter is as likely to be retrograde as direct.

An interesting group of minor planets providing distinct possibilities for capture is known as the Trojan group, the individuals of which are named after the heroes of the Trojan war described in Homer's *Iliad*. The mean distance of a Trojan from the Sun is almost exactly equal to the mean distance of Jupiter from our luminary; accordingly, its orbit is approximately the same as Jupiter's orbit. Several of the Trojans are roughly 60° ahead of Jupiter in its orbit—that is, in the vicinity of A in Fig. 12; the remainder are roughly 60° behind Jupiter —in the vicinity of B in the figure. Owing to the attraction of Jupiter in the peculiar circumstances described the disturbances on the orbit of a Trojan are considerable; as viewed from Jupiter the Trojan (in the first group) describes a roughly elliptical path around A, being nearest to Jupiter at X and farthest at Y. The orbital motion of the Trojan around the Sun is of course a combination of the orbital motion of Jupiter and the motion in the broken curve surrounding A. A similar situation arises in connexion with the Trojans in the neighbourhood of B. Although there is little chance of the immediate capture of any of the known Trojans, yet in the past a minor planet of this type may have approached Jupiter so closely, under the circumstances already indicated for the minor planets in general, that capture ensued. It is thus not beyond the bounds of possibility that the three outermost and retrograde satellites of Jupiter were formerly minor planets, either of the ordinary type or of the Trojan type. The mathematical

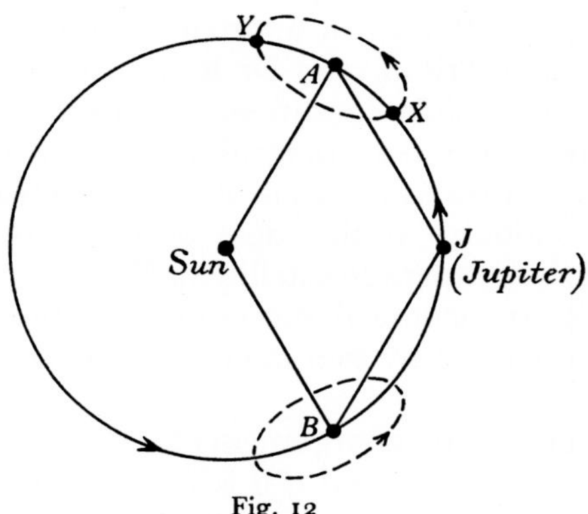

Fig. 12

problem is obviously one of the utmost difficulty and complexity, and it is hardly surprising that the suggestion of satellite capture in the way roughly indicated as it affects Jupiter has not been lifted out of the trough of speculation into the higher levels of mathematical demonstration.

There is one example, however, of a retrograde satellite for which a rather more detailed investigation is possible; this is Triton, the large satellite of Neptune. Triton is the most massive satellite in the Solar System, its mass being about one-seventeenth of the Earth's mass, nearly five times the mass of the Moon and about $2\frac{1}{2}$ times the mass of Ganymede, the next most massive satellite. As previously stated, the details relating to Pluto are not known with any great accuracy; it is, however, believed that Pluto is about as massive as Mars—the mass of Mars is about one-tenth that of the Earth; accordingly, Triton and Pluto are not very dissimilar as regards mass and, probably, in other characteristics. Now the orbit of Pluto is very elongated and a small part of it actually lies inside Neptune's orbit which is very nearly circular with a radius of thirty astronomical units or nearly 2800 million miles. When nearest the Sun, Pluto is about 2750 million miles distant. Although the plane of Pluto's orbit is inclined at an angle of about 15° to the plane of Neptune's orbit, Pluto can approach comparatively closely to Neptune and Triton, and Dr R. A. Lyttleton has shown that under favourable conditions the mutual attractions of all the bodies concerned can result in the capture of Pluto by Neptune and the transformation of Triton's retrograde motion into direct motion. If this does occur in the future, the anomaly of the retrograde satellite of Neptune will no longer exist for our astronomical successors and they will see the planet accompanied by two moons revolving in the normal direction.

But the mathematical argument can be applied to the past as well as to the future and if it is supposed that Triton and Pluto were at one time direct satellites of Neptune moving in orbits such that a close encounter of Triton with Pluto were possible, then the result under suitable circumstances would be the reversal of Triton's direct motion into retrograde motion and the change of status of Pluto from a mere satellite to an independent planet of the solar system. This explanation of the present retrograde motion of Triton is very plausible, particularly in view of the satellite-like dimensions of Pluto, and if it can be substantiated by a more complete mathematical analysis, the apparent anomaly of the retrograde motion of one satellite, at least, would be removed.

It is sometimes believed that the Solar System is like a perfect mechanism working smoothly and changelessly; we are then apt to forget the significant gravitational interactions of its constituent members and to fail to realize that such features as seen at present to be puzzling and at variance with the general uniformity observed can be conceived to have emerged by natural processes from a state, in the remote past, in which non-uniformities may be expected to be absent. It may be added that the resources of the mathematician in working out, in complete detail, the past or future history of the Solar System are considerably circumscribed. The late Professor E. W. Brown, the greatest authority on these matters in the present century, once stated that the theory of planetary motions, so far developed, is adequate to foretell the circumstances of the Solar System with reasonable accuracy for no more than 100 million years; the same interval applies equally to the past history of the planets. This, it will be agreed, is an extraordinary achievement; on the other hand we must not forget that the interval within which the planets are securely held within the bonds of mathematical analysis is but a small part, as we shall see, of the lifetime of the Solar System as an organic unity.

We consider now another uniformity relating to the major planets. Except for Mercury and Pluto, at the nearest and farthest bounds of the Solar System, the eccentricities of the orbits around the Sun are extremely small so that the orbits differ very little from circles. Perhaps we can ignore Pluto as an apparent exception, for its present circumstances are probably the result of the process, just described, involving change of status. In whatever way the planetary system came into being, it is hardly likely that the planetary orbits were nearly circular originally. Although the eccentricities can be modified to some extent through the gravitational interaction of the planets themselves, we must look further afield for an explanation. It is known that if the space within which a planet revolves around the Sun is supposed to be occupied by a cloud of gas or particles, the resistance to motion offered by the cloud has several effects on the orbit, one being the progressive diminution of the eccentricity. It would then seem not unreasonable to suggest that some time in the past the Solar System was immersed in an extensive cloud—or resisting medium, as it is called in general terms—with the result that the elongated orbits of the planets became, in due course, reduced to approximately circular paths. The resisting medium may have been contemporaneous with the birth of the Solar System, as supposed in some theories regarding its origin; or it may have been one of the many interstellar clouds or diffuse nebulae so common in the stellar system, in which event the Sun with its attendant planets must be supposed to have passed through it, like a fleet through a fog. We cannot judge which of the two alternatives is to be favoured, if it is assumed that the rounding of the orbits has actually taken place.

The first alternative, involving as it does the hypothesis of the common ancestry of the resisting medium and the Solar System, makes little contribution to the solution of our problem. The second alternative is equally unhelpful, for it supposes that the Solar System was already in being. In any event we have a plausible explanation of the approximate circularity of the planetary orbits, although this particular uniformity is unable to fortify our argument as to the unitary character of the Solar System.

The chemical composition of the bodies of the Solar System, so far as this has been ascertained, gives us little assistance in supporting or contradicting the thesis relating to the common origin of the planets. It has been shown, as we have seen (p. 53), that about sixty of the ninety-two terrestrial elements exist in the Sun* and that there are very good reasons why the remainder have so far eluded discovery. It is noteworthy that, so far as can be ascertained, the order of abundance of the common metallic elements—iron, aluminium, magnesium and so on—is much the same in the Sun as in the Earth's crust, and that many elements rare terrestrially are also rare in the Sun; it would then seem that the evidence provided by such elements is not inconsistent with the hypothesis that the Earth was at one time part of the Sun. But when we turn to the

* More accurately, 'in the Sun's outer atmospheric levels'.

consideration of the most abundant element in the Sun—namely, hydrogen which forms, as it will be recollected, about one-third of the solar mass—we find a different state of affairs, for hydrogen is a comparatively rare element in the Earth. If the hypothesis is correct, it is reasonable to suppose that the Earth originally contained a substantial proportion of hydrogen. In earlier pages (pp. 45–49) we have discussed at length the capacity of a planet to retain an atmosphere and have shown how utterly incapable the Earth is to retain hydrogen in its atmosphere at the present time and even more so in the distant past when, as the inexorable laws of cooling tell us, it was very much hotter, perhaps molten or even gaseous. The present state of the Earth as regards its hydrogen content is thus what would undoubtedly be expected on the supposition of its solar parentage.

When we turn to the other planets we find that spectroscopic exploration can tell us comparatively little about the chemical constitution of these bodies. We are aware of the presence of a small number of the elements in the atmospheres of the more massive planets but, beyond that, our knowledge comes to a dead stop. Although, owing to a variety of circumstances, the spectroscopic method of analysis may fail in the future to establish fully the chemical composition of the planets, it is perhaps not unreasonable to surmise that the planets—and the satellites—are on the whole not dissimilar to the Earth in this respect, except in the relative abundance of the various elements.

Spectroscopic investigations of the stars reveal that many stars are very much like the Sun in chemical constitution, both as regards the elements detected and also as regards their relative abundance. Of course there are many stars that are not so self-revealing. For example, the spectra of the hottest stars show little more than that hydrogen and helium are present in the atmosphere in great abundance; it is not to be supposed, however, that such stars are composed entirely of these two gases, since the physical conditions in the stellar atmospheres are such that the great majority of the other elements would fail to record their familiar spectral messages. In searching for the parent of the planets we have now widened the field to an almost unlimited extent, for it is not beyond the bounds of possibility that, through the gravitational interaction of a star and the Sun, the planets have been formed out of the stellar material and have been acquired by the Sun during the process. Indeed, this particular view is the starting-point of at least one theory of the origin of the planetary system.

In this concluding section we briefly summarize such progress as we have made in the quest for the answer to the first of our three principal questions 'Whence?' We first repeat that the idea cannot be seriously entertained of the piecemeal growth of the solar family by fortuitous capture, for it is inconceivable that the uniformities of orbital and rotational motions could have been established in this way. The planetary system must then have come into existence as the product of some single cosmic process. Our investigations on the chemical constitution of the Earth, the planets and the Sun are unable to specify definitely the exact parentage of the so-called solar family. It is not unlikely that the origin of the planets is to be looked for in the Sun itself, although the possibility must not be overlooked that some unknown star, now roaming in the depths of interstellar space, is the real material parent. In due course we shall consider the various suggestions of the process whereby the Sun, either by its own exertions so to speak, or in collaboration with another star, produced the orderly system of planets.

Part II

WHEN?

CHAPTER VI

THE GEOLOGICAL RECORD AND THE CONFLICT WITH PHYSICAL SCIENCE

In this chapter we begin our attempts to answer the question: When was the Earth formed? restricting our discussion here mainly to purely geological arguments together with the counter-arguments of physical science antecedent to the discovery of radioactivity at the end of last century. We must first of all make clear what is involved in this question relating to the age of the Earth. The Earth is of course composed of a multitude of atoms of the elements, but our present inquiry does not extend back in time to the possible building-up of atoms from still more elementary particles. We refer to the age of the Earth as the interval of time during which the Earth has existed as an independent member of the solar family, and more particularly as the interval since its solid crust was formed.

The records of intellectual activity in historic times show that the problem of the Earth's age has been a continual source of speculation mainly bound up with religious beliefs and aspirations. It is perhaps sufficient to mention the maximum and minimum 'ages' attributed to the Earth. The first is inherent in Hinduism; according to the *Upanishads*, man's soul has passed through—and will continue to pass through—an endless series of changes, with the implication that terrestrial time is infinite in the past and infinite in the future. The second is associated with the famous date, 4004 B.C., of the Creation of the Earth and the Firmament as estimated from Hebrew chronology by Archbishop Ussher in the seventeenth century.* It is only within comparatively recent times that

* Later theologians were not quite satisfied with an accuracy that specified only the *year* of creation; according to them the moment of the

the establishment of natural laws and the recognition of the effects of familiar processes have led to a reasonably reliable estimate of the Earth's age, not however with the kind of precision implied in Ussher's date of creation.

At the beginning of the nineteenth century geologists were divided into two camps, the 'catastrophists' and the 'uniformitarians'. The former, familiar as they were with the immense changes produced on the Earth's surface by volcanoes, earthquakes and vast floods, saw in the rugged mountain-ranges and the deep depressions of sea and land evidence of vast cataclysmic episodes in the past. From classical times onwards the remains of sea-organisms were discovered on dry land hundreds of feet above sea-level, evidently unmistakable support for the dogmas of 'catastrophism', for how otherwise, it was asked, could a sea-bottom be elevated above sea-level except by a violent convulsion of Nature?

The uniformitarians, on the other hand, explained the characteristics of the Earth's crust in terms of processes that were going on incessantly even if only slowly. They saw how rocks decayed and how the solid matter of the uplands was gradually transported by rivers and glaciers to lower levels and eventually to the oceans, there to form sedimentary strata some of which, by slow readjustment of the crust under the strains and stresses of changing loads, became elevated to form the continental areas. The postulation of the slow processes of denudation and sedimentation and the recognition of the immense thicknesses of the sedimentary rocks implied a vast geological time-scale compared with which the interval since 4004 B.C. seemed but the tick of a clock. James Hutton, the originator of the uniformitarian theory and the first great British geologist, went so far as to declare, towards the end of the eighteenth century, that 'in the economy of the world

Earth's creation was 9 o'clock on the morning of 23 October, 4004 B.C., the hour presumably being defined with reference to our present system of specifying Greenwich Mean Time.

I can find no traces of a beginning, no prospect of an end'. More modern views, as we shall see, do not subscribe to the boundless existence of the Earth in the past but, nevertheless, the period of time required for geological evolution is almost unbelievably immense.

The study of fossils in sedimentary rocks furnished a powerful means for advancing geological knowledge. A fossil is either the remains of the whole or part of a prehistoric organism, or simply evidence of the past existence of such an organism. The best conditions for the preservation of fossils are afforded by shallow seas; the lifeless bodies of whatever forms of sea-life there may be are soon covered by sand and mud which slowly, by accretion, form eventually a sedimentary stratum perhaps several thousand feet thick; further, the bodies of land animals may be transported by rivers to the sea-beds, there to leave a record of their previous existence.

Palaeontology (the science of fossils) divides fossils into three main groups. The first type refers to the actual remains of an organism; these may be skeletons imprisoned within the rocks or, more frequently, the soft parts of organisms almost miraculously preserved or, occasionally, remains of both skeletal and soft parts; for example, flies are often found preserved in amber—the solidified resin peculiar to certain types of trees—and the remains of mammoths, almost wholly preserved, are frequently found in the ice-sheets of northern Siberia. In the second type of fossil the actual substance of the original organism has been replaced by some mineral substance such as silica or calcite; in some fossils of this type the actual structure is preserved as in petrified wood, in others only the shape of the organism is preserved, while in others only an impression on the rock testifies to the previous existence of the organism. The third type of fossil consists of tracks made by the feet of animals such as the dinosaurs in the age of the terrestrial monsters, or trails or tunnels made by such organisms as worms.

Although the existence of fossils has been known since

classical times, it was only at the beginning of last century that an intensive and systematic study as to their geological and biological significance was set on foot. The French scientist Cuvier was the first to realize that the fossil bones discovered in the rocks near Paris were wholly different in structure and size from the bones of living animals; here was evidence of extinct orders of the animal kingdom. Moreover, he found that each stratum had its own particular species of fossils, from which he concluded that at intervals, corresponding to the beginnings of the formation of the different strata, immense cataclysms had occurred which resulted in the complete extinction of all organic species in the regions affected; once the particular crustal commotion had subsided the Earth was then supposed to be repopulated by new species. This was the doctrine of catastrophism.

About the same time, at the beginning of last century, Lamarck was studying the fossils of invertebrates (boneless organisms). His conclusion was that, although each stratum had on the whole its own peculiar series of organisms, yet some types persisted through several strata while others appeared to show a progressive structural and functional development through successive geological ages. He was disposed to affirm that such organic changes were evidence of the capability of living things to adjust themselves to changing environment and physical conditions: this, in brief, is the theory of Lamarckism which was supplanted in the second half of last century by Darwin's theory of 'natural selection' as the guiding principle of evolutionary change.

A further step in the deciphering of the Earth's crustal history was taken in the first decades of last century by William Smith, a land surveyor and civil engineer, who was born at Churchill, near Oxford. He found that certain fossils occurred only in certain rocks, from which he inferred that all rocks, wherever found on the Earth, which contained a particular fossil are of the same geological age. Adopting the principle that the most primitive forms of plant and animal life are found in the oldest rocks and that progressive development in the structure and functions of terrestrial organisms occur at successive epochs, geologists have been enabled to arrange the principal geological formations in order of age, with specific information as to the appearance and disappearance of various distinctive species. The following table gives the geological eras in the first column and the recognized subdivisions in the second column. The ages of the rocks are inserted in the table for reference; we shall show later how these ages have been obtained. The most distant era is simply denoted here by the term 'Pre-Cambrian' which is variously subdivided by geologists.

TABLE V. *Time-scale of the Geological Subdivisions*

Era	Subdivision	Age (in millions of years)
CAINOZOIC (Age of Mammals)	Pleistocene	5
	Pliocene	15
	Miocene	30
	Oligocene	40
	Eocene	70
MESOZOIC (Age of Reptiles)	Cretaceous	110
	Jurassic	140
	Triassic	190
PALAEOZOIC (Ancient Life)	Permian	220
	Carboniferous	280
	Devonian	320
	Silurian	340
	Ordovician	390
	Cambrian	500
PRE-CAMBRIAN	—	Up to 1800

Let us briefly describe the pageant of terrestrial life as revealed in the records of the rocks. When life first appeared is not certain. In the latest of the Pre-Cambrian rocks primitive sea-plants such as seaweed and sponges are found, while the trails and burrows of worms reveal the existence of the earliest forms of animal life. In the Cambrian subdivision of the Palaeozoic era life seems to have been abundant; the presumed ancestors of the principal types of invertebrate animals appeared and these progressed in number and organic development as the ages advanced. In the Silurian subdivision fishes and the first of the land-animals appeared; in the Devonian subdivision fishes were abundant (this subdivision is sometimes referred to as the 'age of fishes'), while land-plants were widespread. In the Carboniferous subdivision, rapidly growing and luxuriant vegetation covered the swamps and after compression by later sediments gave rise to the coal seams which have made modern industry possible; at this time, too, flying insects first appeared. The Permian subdivision of the Palaeozoic era saw the Earth subjected to severe glaciation, with the result that many species became extinct. In the Mesozoic era reptiles swarmed in the sea and on land; some, the pterodactyls, even took to the air; monsters roamed everywhere the largest of which, the diplodocus, was nearly 100 feet long; the development of birds went on apace. And then something of a zoological catastrophe occurred, for the mammoths completely disappeared whether as a result of a crustal cataclysm or of the inability of these creatures to withstand the violent assaults of more vigorous species or of the insidious attacks of bacteria is not known. In the Cainozoic era mammals appeared and prospered, culminating—perhaps a million years ago—in Man. It must not be inferred from this brief sketch that the geological record is complete even since the beginning of the Cambrian subdivision; nevertheless, although gaps still remain to be filled in, geology gives us a broad and coherent picture of the succession of plant and animal species from the very distant past to the present time.

With the evidence of the slow evolution of the Earth's crust and of terrestrial life it is not surprising that the geologists and biologists of the nineteenth-century uniformitarian school instinctively demanded an almost inconceivably long time-scale of thousands of millions of years. It is true that, in those days, they had no accurate estimates of the ages of the stratified deposits, for reliable methods of chronology were only developed fully in the present century. However, the insistence on the long geological time-scale brought geology and biology into sharp conflict, in the second half of last century, with the apparently incontrovertible arguments advanced by Lord Kelvin as to the age of the Earth; by three different lines of attack based on the principles of physics then established, Kelvin showed that the Earth could not have existed as an independent planet for more than roughly 100 million years. Thus was joined one of the great scientific contests of the Victorian era, which was only settled in favour of the geologists by a new discovery in the realm of physics—radioactivity—at the end of last century, although within the framework of contemporary knowledge Kelvin's arguments were almost completely irrefutable.

In addition to their intrinsic interest in the historical controversy, Kelvin's three ways of tackling the problem of the Earth's age have distinctive features fundamental in present-day researches. We deal with them in turn. The first concerned the implications of frictional forces in world-economy. At the beginning of the nineteenth century the labours of the great mathematical astronomers on the motions of the planets appeared to have achieved one significant result, namely, the proof of the stability of the Solar System through long stretches of past and future time; the planetary system was regarded as a vast machine which, once put into motion, continued uninterruptedly to perform its functions without any suggestion of breaking down and whose state at any time, in the reasonably remote past or future, could be accurately determined from mathematical formulae. This deterministic view is summed up in the well-known words of the great French astronomer Laplace: 'We ought to regard the present state of the universe as the effect of its antecedent state and as the cause of the state which is to follow. An intelligent being who at a given instant knew all the forces animating Nature and the relative positions of the objects within it would, if his intelligence were sufficiently capacious to analyse these data, include in a single formula the movements of the largest bodies of the universe and those of its lightest atom. Nothing would be uncertain for him; the future as well as the past would be present to his eyes.'

It is clear that this evidence from astronomy fitted in satisfactorily with the uniformitarian view, stressing the necessity for almost endless time in past geological and biological development. Just as Hutton (p. 102) could find no traces of a beginning, no prospect of an end in the phenomena of geology and biology, so the later apologists of uniformitarianism accepted the conclusions of astronomy that, in Playfair's words, 'in the planetary motions where geometry has carried the eye so far both into the future and the past, we discover no mark either of the commencement or the termination of the present order'. But the great mathematical researches on planetary motions, to which we have alluded, took no cognizance of the consequences of possible frictional forces on planetary motions except to show that if such forces operated their effects were so minute as not to be capable of detection from the observations made within historic time. This was the position up to 1853 when J. C. Adams, the joint discoverer of the planet Neptune, arrived convincingly at a contrary conclusion. The frictional forces are no doubt minute but not so small as to fail to leave their mark on celestial motions. The problem, of great intrinsic interest, is relevant in our discussion of the Earth's age and we set out the main points, remarking at the outset that the Earth and the Moon are the two bodies that immediately concern us.

The rotation of the Earth furnishes us with the basis for the measurement of time; the rotation is in the direction from west to east. We are not, however, conscious directly of this rotation—for the Earth seems stationary—and consequently we get the impression that the firmament rotates in the opposite direction; the stars thus appear to us to traverse the heavens from east to west. An accurate instrument, known as the meridian circle, enables an astronomer to note the instant when a star crosses his meridian; according to circumstances the star is then either due north or due south. When the star next crosses the meridian a sidereal day is said to have elapsed. Disregarding some niceties of adjustment that need not concern us here, we conclude that the sidereal day is simply the period of rotation of the Earth about its axis; this is the basis of time-measurement. To measure sidereal time the astronomer is equipped with a 'sidereal clock', the 24 hours which its dial displays corresponding to the sidereal day, with minute and second hands performing the usual subdivisional functions. When we said earlier that the observer was able to note the instant at which a star crosses his meridian, it is to be understood that his observation is made with reference to such a sidereal clock. We may further suppose that the clock keeps sidereal time accurately; actually, of course, no mechanism is perfect, but observations of representative stars enable the astronomer to keep a check on the performance of his clock and to say what the correct reading at any instant ought to be.

One additional remark about the setting of the sidereal clock may be made. When a particular star is on his meridian the clock is supposed to be set to the reading $0^h\ 0^m\ 0^s$; in practice this star is fictitious, being a particular point of the heavens called the 'First Point of Aries'; the principle is, however, fundamentally the same as if the 'star' were real. It may be added that mean solar time, according to which our ordinary daily activities are regulated, is intimately connected with sidereal time. Until comparatively recent times it was believed that the rotation of the Earth about its axis was uniform and constant: thus, it was supposed that the length of the sidereal day as measured by a perfect sidereal clock would be precisely 24 hours—no more and no less—from day to day and year to year; in other words, the rotating Earth was itself the perfect clock. It is now found that the Earth's rotation is gradually slowing down with the result that the sidereal day is increasing in length, but only at such a minute rate that the effect is not of the slightest consequence in our, or the astronomer's, ordinary affairs. The effect is, however, important in the cosmological problem with which we are more immediately concerned here and its discovery is one of the fascinating stories of modern astronomy. The observed motion of the Moon provides the clue.

If the Earth-Moon system were completely removed from the gravitational influence of all other bodies, then, with one or two further qualifications of a comparatively minor sort, the Moon would appear to describe an ellipse around the Earth's centre. But the Sun and the planets are all attracting the Earth and Moon, and each other, and the problem of determining accurately the Moon's path relative to the Earth becomes one of the greatest complexity. However, all we need consider here is the fact that successive generations of mathematical astronomers have tackled the problem with ever-increasing attention to detail. Also, it must be added, the path of the Earth around the Sun has been investigated with the greatest possible refinement. It is thus possible to calculate for any instant, first, the Moon's position in the sky and, second, its position in relation to that of the Sun. Just as a particular position on the Earth's surface is specified by latitude and longitude, so the Moon's position in the sky is specified by two similar co-ordinates, celestial latitude and celestial longitude, the latter being measured along the ecliptic. Let us simplify the problem by ignoring the Moon's celestial latitude, that is, by assuming that the Moon moves in the sky along the ecliptic; this motion is eastward with reference to the background of the stars and the Moon's longitude —dropping the adjective 'celestial'—increases through 360° as the Moon completes its circuit of the heavens.*

* The parallelism between celestial longitude and terrestrial longitude would be complete if the latter were measured eastward from the Greenwich meridian through 360°; however, it is more convenient for the ordinary purposes of terrestrial relationships to measure the terrestrial longitude eastward up to 180° and westward up to 180°.

At any instant, then, the Moon's longitude and its position relative to the Sun can be calculated; in particular, we can calculate accurately the times at which eclipses of the Sun by the Moon, or of the Moon by the Earth, occurred in the past or will occur in the future, all on the implied assumption that the rotating Earth is a perfect time-keeper. As we have reliable records of several ancient eclipses it is thus possible to check theory with observation. When the records were examined, in the middle of the eighteenth century, it was found that the eclipses had actually occurred at times earlier than those obtained by calculation; for example, the discrepancy for one of the earliest eclipses investigated, an eclipse of the Moon observed at Babylon in 721 B.C., was found to be nearly $3\frac{1}{2}$ hours; later eclipses, it may be added, gave successively smaller discrepancies. The inference drawn from the discussion of these eclipse results was that the Moon seemed to be moving more rapidly in its orbit round the Earth in the eighteenth century than in previous centuries and that, further, its average angular rate of motion in longitude—we refer to this simply as the *mean motion*—seemed to increase at a steady rate per century; expressed in a more technical way, the ancient eclipses indicated that the Moon's longitude was subject to an unexplained acceleration, known briefly as the Moon's *secular acceleration.*

The effect of the secular acceleration on the Moon's longitude may be exemplified in one or two cases. Suppose that the Moon's longitude is determined accurately from observations at some time in 1700; from gravitational theory its longitude precisely one century later can be calculated on the assumption that the Earth's rotation is constant, and when this is compared with the observed longitude it is found that the latter is about 10 seconds of arc greater; two centuries later, that is in 1900, the discrepancy will be 40 seconds of arc and three centuries later—that is in 2000—the discrepancy may be expected to be 90 seconds of arc; going backwards in time to the Babylonian eclipse of 721 B.C., the discrepancy will be 1° 38′ nearly—in the sense now that the observed longitude is less by this amount than the calculated amount; since the Moon moves about $\frac{1}{2}$° in longitude per hour, the discrepancy of 1° 38′ gives a discrepancy of a little over 3 hours between the calculated and observed times of the eclipse.

The efforts of the great mathematical astronomers of the eighteenth century were now directed to elucidate the mystery of the secular acceleration, the apparent reality of which could not be seriously questioned. There were two possibilities: either the mathematical theory of the Moon's motion according to Newtonian principles was inexact or some unknown physical process was operating. If we consider the second alternative first, the secular acceleration could be easily explained if the Earth's angular speed of rotation is slowly diminishing, this implying that the day is gradually lengthening. It had been suggested as far back as the time of Kant that the terrestrial tides must produce a braking action on the Earth's rotation, but until very recent times there was no successful attempt to determine the numerical effect of such an action and to show that it accounted fully for the secular acceleration. However, the first alternative had first to be disposed of and the theory of the Moon's motion was critically re-examined by the leading mathematical astronomers of the eighteenth century without any enlightenment on the phenomenon until Laplace, in 1787, announced the solution of the mystery: an apparently trivial detail in the mathematical theory had been overlooked and when its implications were included the secular acceleration was completely and exactly accounted for—in other words, the Moon's position at any instant as calculated from the revised theory agreed with the observed position at that instant. Thus one of the outstanding problems of astronomy appeared to be settled satisfactorily and finally; further, it now seemed certain that the braking effect of the tides on the Earth's rotation must be insignificant, from which it was concluded that the rotating Earth was, in fact, a perfect time-keeper.

And so the matter rested until 1853 when, as already mentioned, J. C. Adams, the joint discoverer of the planet Neptune, showed that a subtle point in the lunar theory had been missed by Laplace; the theoretical coefficient of the secular acceleration was now reduced to $5\frac{3}{4}$ seconds of arc, just a little more than half the coefficient* demanded by the eclipse records. Adams's calculation was confirmed by the eminent French astronomer, Delaunay, while four other distinguished astronomers obtained four distinct results, differing widely amongst themselves and from the $5\frac{3}{4}$ seconds of arc found by Adams and Delaunay; the only unanimity discernible was that each of the astronomers (regarding Adams and Delaunay as one) believed that all the others were wrong! It is now universally conceded that Adams and Delaunay were right; in addition, the unexplained part of the secular acceleration, now amounting to about 5 seconds of arc, must be attributed to a non-gravitational source such as the tidal retardation of the Earth's rotation.

Before dealing with Kelvin's objection to the long time-scale of the uniformitarians we bring the story of the Moon's secular acceleration up to date. Although Kelvin used as his argument the effect of the friction produced by oceanic tides it is now known that this is wholly insufficient to yield the quantitative result demanded by the outstanding discrepancy (5 seconds of arc) between the calculated and observed values of the secular acceleration. In 1920 Sir Geoffrey Taylor showed the efficacy of tidal friction in the shallow waters of the Irish Sea and later investigations by Professor H. Jeffreys, dealing with tidal currents in various parts of the globe, confirmed the slowing down of the Earth's rotation by substantially the amount required by the ancient eclipses. Owing to tidal friction the length of the day is increasing at present by about two-thousandths of a second (of time) per century—a small rate, but with important consequences, as we shall see later (Chapter VIII) when we envisage the past history of

* The modern value of the coefficient is 11 seconds of arc.

the Earth over the immense periods of time with which we are concerned.

We now consider Lord Kelvin's first argument against the long time-scale demanded by the uniformitarian geologists and biologists. As we have seen, he attributed Adams's result to the slow retardation of the Earth's rotation brought about by frictional forces. Going back 1000 million years he calculated that the Earth must have been rotating nearly 15% more rapidly than at present. If the Earth were fluid at that time the theory of rotating fluid masses showed that its shape must have been much more flattened than it is at present, with the polar diameter very much smaller than the equatorial diameter. If the crust were formed about this time, the marked disparity between the polar and equatorial diameters should have persisted, he argued, to the present time; the fact that the polar diameter is now just a little less than the equatorial diameter appeared to rule out of court the assumption that the Earth solidified 1000 million years ago; but if it is assumed that the Earth solidified only 100 million years ago, the shape of the Earth would be substantially that at present. Thus, Kelvin arrived at the conviction that the Earth's age, defined with reference to the time at which its crust was formed, was unlikely to exceed 100 million years. The argument, of course, implies that the original crust retained its shape and structure without substantial modification, whatever its age, until the present day; but the geological processes of crustal readjustment and continental uplifts and subsidences, now familiar from more recent investigations, introduce doubts into Kelvin's main line of argument.

Kelvin's second argument against the long time-scale was based on his calculation of the 'age of the Sun', for it is agreed that, whatever the Sun's age may be, that of the Earth is not likely to be greater. The investigation involved considerations as to how the Sun maintained its outpouring of heat over vast intervals of time. The geological and biological record was emphatic in asserting that uniformitarianism also applied to the Sun—in other words, that the Sun poured out heat at much the same rate in the distant past as at present. In Kelvin's time the doctrine of the conservation of energy, to which he himself made notable contributions, had been securely established. Heat is a form of energy and if the Sun is continuously pouring forth heat into space it must have some source of replenishment of energy to maintain the output.

Kelvin at first suggested that the Sun's heat was derived from the kinetic energy of meteors falling on the Sun. We have already seen that when meteors enter the Earth's atmosphere they are vaporized almost instantaneously; in this case, the energy of motion of a meteor is converted mainly into heat and light energy. In the same way meteors falling into the Sun would undoubtedly go some way, at least, to make up for the wastage of the solar heat by radiation. However, this idea was abandoned later because of the many difficulties in accounting for the continuous and multitudinous streams of meteors required to maintain the Sun as a going concern. Instead, Kelvin adopted the suggestion originally due to Helmholtz that the heat expended in radiation was provided by the diminution of the potential energy of the solar matter arising from the gradual contraction of the Sun through the mutual gravitational attraction of its parts. The rate at which the Sun is expending energy in the form of heat is known, and the amount of energy gained by contraction since the Sun was a greatly extended globe can be calculated if we make some reasonable assumption as to the way the density of the solar matter at present varies with distance from the Sun's centre. With the data at his disposal and realizing that his estimates were necessarily somewhat uncertain, Kelvin concluded that the Sun may have illuminated the Earth for as long as 100 million years and was convinced that an age of 500 million years was wholly impossible. Modern observational data and a more detailed mathematical treatment of the problem of a contracting gaseous sphere completely justify Kelvin's conclusion and are more precise in fixing the Sun's age at not more than 50 million years.

It is to be understood that in all these calculations, Kelvin's and the modern, the Sun is assumed to be drawing on contractional energy only to maintain its radiation of heat; since Kelvin's time new sources of energy have been discovered which, as we shall see in a subsequent chapter, greatly extend the Sun's life. But in the nineteenth century Kelvin's argument was irrefutable; the geologists were completely discomfited, for they had no adequate reply to the remorseless logic of mathematics.

Kelvin's third line of attack was equally disconcerting. As already stated (p. 58) the Earth's temperature increases with depth below the surface, the temperature-gradient close to the surface being about 1° C. for 80 feet in depth; heat is thus flowing outwards from the Earth's interior and is eventually radiated from the surface into space. The Earth, then, is a slowly cooling globe having been, presumably, very much hotter in the past; in making these assertions we imply that there is no process, within the Earth, of producing heat to counterbalance the loss by eventual radiation into space. Assuming that the Earth was at one time a molten globe with a uniform temperature of 4500° C. and using what was then known about the physical properties of rocks at high temperatures, Kelvin solved the problem of finding how long such a globe would take to cool to provide the observed temperature-gradient already mentioned. The answer was found to be just under 100 million years.

Later, when more was known about the physical properties of rocks, Kelvin was of the opinion that the assumed initial temperature of 4500° C. was excessive;* the effect was to reduce considerably the Earth's age, as measured from its

* The melting-point of silica (one of the most refractory substances and a chief constituent of the rocks) under normal pressure is now known to be just about half of this temperature.

molten state, to about 25 million years. It should be added that the possibility was considered that the outflow of heat from the Earth might be derived, wholly or in part, as a result of chemical combinations within its interior, in which event the Earth's age must be extended beyond the limit mentioned. However, such a hypothesis was shown to have little bearing in increasing the Earth's age substantially because the most energetic chemical action then known, even on the assumption that it involved the whole mass of the Earth, could not produce enough heat to prolong the Earth's age by more than a relatively insignificant amount.

The controversy between the physicists, led by Lord Kelvin, and the geologists lasted until the end of last century. Many of the latter resolutely ignored what seemed to be the unimpeachable evidence of physics, taking up the position that geological problems were the concern of geologists only. Their attitude is well summarized in the words of T. H. Huxley: 'I do not suppose that at the present day any geologist would be found to maintain absolute uniformitarianism, to deny that the rapidity of the rotation of the Earth *may* be diminishing, that the Sun *may* be waxing dim or that the Earth itself *may* be cooling. Most of us, I suspect, are Gallios "who care for none of these things" being of opinion that, true or fictitious, they have made no practical difference to the Earth during the period of which a record is preserved in stratified deposits.' This Victorian attitude—that each science operated within its own water-tight compartments, as it were, brooking no interference from other sciences—is fortunately now a thing of the past. Geology now owes a great debt to physics, for the impasse brought about by Kelvin's authoritative researches was only ended by revolutionary physical discoveries made towards the close of the nineteenth century.

We now turn to the geological arguments for the long timescale demanded by the uniformitarians. There were two principal methods suggested, one depending on the rate of increase of the salinity of the oceans, and the second on the rate at which the sedimentary deposits are laid down. The principles of applying each method are similar and they can be illustrated very clearly with reference to the Dead Sea. As is well known, the Dead Sea is a vast natural reservoir, about $\frac{1}{4}$ mile below the level of the Mediterranean Sea, into which run the waters of the River Jordan but from which there is no exit. The Jordan carries in solution to the Dead Sea several chemical salts—principally the chlorides of magnesium, sodium, calcium and potassium, with one or two other compounds—and through evaporation, extending over many thousands of years, the concentration of these chemicals in solution has gone on increasing. It is reliably estimated that the amount of potassium chloride in the Dead Sea at present is about 2000 million tons; also, the amount carried in solution daily by the Jordan is about 100 tons. If the daily addition of potassium chloride has been constant, an easy calculation shows that the process must have been going on for about 55,000 years. We thus get some clue to the age of the geographical and geological features presented to us by the Dead Sea. It may be added that the rich store of chemicals in the Dead Sea is now being exploited commercially; it is estimated that about half the potash used in fertilizers for British agriculture during the second World War came from this source.

In a similar way the age of the oceans can be calculated. It was estimated by Joly in 1899 that the amount of sodium (mainly in the form of common salt) carried into the oceans by all the rivers of the globe amounted to 156 million tons annually; the total amount in the oceans was estimated to be 12,600 million tons. If the annual addition is assumed to have been constant throughout the past, a simple division gives 81 million years as the age of the oceans—a rather depressing result to geologists, as it appeared to vindicate Kelvin's general conclusion. However, later calculations raised the age of the oceans to 330 million years but even this figure is now regarded by geologists as greatly underestimated when various factors, now better understood and more accurately calculable, are taken into consideration. The general consensus of opinion now seems to be that the age of the oceans indicated by this method must be at least 1500 million years. The oceans, of course, came into existence when the Earth had cooled sufficiently from its initial molten state; on the cosmic time-scale this interval was probably short, but in any event we conclude that the Earth's age must be at least 1500 million years.

The second geological method is much more difficult to apply. It involves estimating, first, the total amount of solid material transported annually to the oceans by rivers and glaciers and, second, the total thickness of sedimentary rocks. According to Professor Arthur Holmes the Thames annually carries to the sea between 1 and 2 million tons of solid material; for the whole globe the total amount of sediment annually removed to the seas and oceans reaches the immense total of 6000 million tons. Further, the maximum thickness of sedimentary rocks is found to be nearly 70 miles. It is impossible in the absence of detailed information, to estimate with any approach to accuracy the rate at which these deposits were formed; however, the only conclusion that can be reached is that the period of time during which the process of sedimentation has been operating is very much larger than Kelvin's value for the age of the Earth.

It may seem disappointing that purely geological methods are unable to give greater precision to the age of the Earth, but we shall see in the next chapter how geology was presented with an entirely new method, the results of which unambiguously supported those who had stubbornly placed their faith in a time-scale fifty or a hundred times larger than that emerging from Kelvin's researches.

CHAPTER VII

RADIOACTIVITY TO THE RESCUE

THE discovery of radioactivity in 1896 put a new complexion on the controversy as to the Earth's age which we have described in the previous chapter. In radioactive processes heat is generated spontaneously—we amplify this rather bald statement later—and accordingly we have in the Earth (and also in the Sun, if radioactive substances exist there) a source of heat not envisaged in Kelvin's arguments: as these could no longer be maintained the ages of the Earth and Sun must be much greater than those which he calculated. But radioactivity did more than put Kelvin's calculations out of court; it provided a precise method for determining the ages of the rocks, thus yielding a reliable clue to the age of the Earth itself, and it inaugurated a new era of physical discoveries, relating to atomic structure, unprecedented in the history of science. As the main conceptions of present-day physics are germane to our general theme, we consider in some detail the constitution of the atom, relegating to what may seem a postscript the application of the new ideas to the determination of the Earth's age.

In the fifth century B.C. the atomic theory was first stated by Leucippus and Democritus in the form that matter was ultimately resolvable into discrete entities called *atoms* which could not be further subdivided. Against this was opposed the view held and propagated by Aristotle that matter was continuous and capable of infinite subdivision; this latter view persisted for nearly 2000 years. Both these theories of the nature of matter were philosophical speculations, untested by the application of what we now call the scientific method.

The atomic hypothesis was revived at the beginning of the nineteenth century by the English chemist, John Dalton, basing his arguments on the laws of chemical combination. For example, carbon combines with oxygen in two different ways: in the first, as when a small piece of wood is set alight, twelve parts by weight of carbon combine with thirty-two parts by weight of oxygen to form the familiar gas carbon dioxide; in the second, when the supply of oxygen is severely restricted as in the heart of a coal fire, twelve parts by weight of carbon combine with sixteen parts by weight of oxygen to produce the poisonous gas carbon monoxide; it may be added that on the outskirts of the coal fire the carbon monoxide soon encounters a plentiful supply of oxygen with which it combines, burning with a blue flame, to form carbon dioxide. It is clear that if the same weight of carbon is involved in the two instances, the weight of oxygen required to form carbon dioxide is twice that required to produce carbon monoxide; further if, as Dalton supposed, each of the elements carbon and oxygen consisted of identical atoms, indivisible and indestructible, it follows that the number of atoms of oxygen combining with a given number of atoms of carbon to form carbon dioxide is twice the number of atoms of oxygen combining with this given number of carbon atoms to form carbon monoxide. By a comprehensive study of similar chemical combinations it was concluded that one atom of carbon combines with two atoms of oxygen to form one molecule of carbon dioxide (a molecule, it may be remembered, is a combination of two or more atoms) and that one atom of carbon combines with one atom of oxygen to form one molecule of carbon monoxide; further, the relative weights of carbon and oxygen atoms must be in the ratio 12 to 16. In the formation of water from hydrogen and oxygen, two parts by weight of hydrogen combine with sixteen parts by weight of oxygen and as it was further deduced that the molecule of water consists of two atoms of hydrogen united to one atom of oxygen, the relative weights of hydrogen and oxygen atoms are in the ratio 1 to 16. If we take the weight of the hydrogen atom as unit, hydrogen being the lightest of all the elements, the weight of the oxygen atom is 16 (this is called the *atomic weight*, as originally defined); further, the weight of the carbon atom on this basis is 12 or, in the usual phraseology, the atomic weight of carbon is 12. The chemical elements are ninety-two in number, ranging from hydrogen the lightest, with atomic weight 1, to uranium the heaviest with atomic weight 238 approximately.

The fact that the atomic weights of the lighter elements were found to be whole numbers, within the limits of accuracy then attainable, suggested to Prout in 1815 the hypothesis that the atoms of the elements were built up out of hydrogen atoms; for example, the oxygen atom (atomic weight 16) was supposed to be a tightly bound aggregation of sixteen hydrogen atoms, capable of maintaining its identity in all observable chemical combinations. However, this hypothesis had to be abandoned when exceptions to the whole-number rule were brought to light; for example, the atomic weight of chlorine was found to be about $35\frac{1}{2}$, of copper about $63\frac{3}{5}$, and so on, and consequently it was absurd to suggest that an atom of chlorine consisted of thirty-five hydrogen atoms *and half of a hydrogen atom* when the word 'atom' implied that the latter could not be subdivided. Moreover, when precise methods of deriving atomic weights were fully developed the whole-number rule was not accurately obeyed even by the lighter elements. It is now the practice to take the atomic weight of oxygen to be 16 precisely* and on this basis the atomic weight of hydrogen is 1·008, of helium (a gas and the second lightest of the elements) 4·003, of carbon 12·01 and so on; it is evident that Prout's hypothesis is not accurately, although very nearly, satisfied in the instances just mentioned. Despite its early failure to establish hydrogen atoms as the bricks out of which all matter is built, Prout's hypothesis has been restored—in a modified form, as we shall see—in the modern conception of the structure of matter.

* More accurately, this atomic weight refers to the most abundant isotope of oxygen (see p. 124 for definition of isotope).

Towards the end of the last century it became increasingly evident that matter was much more mysterious than the earlier physicists and chemists had imagined. When an electric discharge was passed through a glass tube (Fig. 13) containing a gas at low pressure various phenomena were observed according to the pressure; the one with which we are concerned occurred when the pressure was reduced to an extremely low value; it was then found that a stream of rapidly

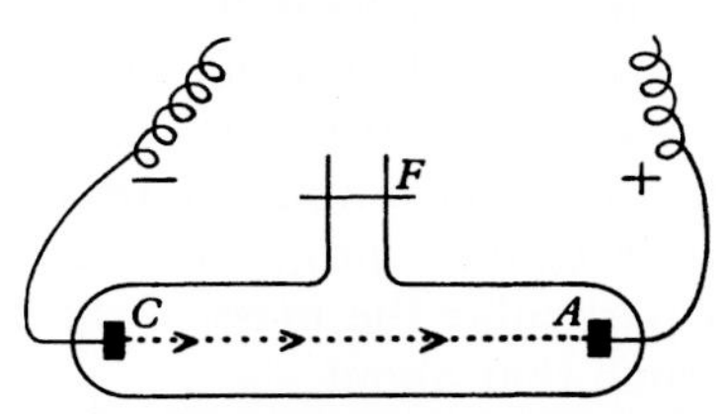

Fig. 13. *A* and *C* are metal disks (electrodes) within the glass tube, which can be connected by wires, sealed in the glass, to a high-tension battery; *F* is a tube connected to an evacuating machine by means of which the gas pressure can be reduced

moving corpuscles passed from the cathode *C* (the negative electrode) to the anode *A* (the positive electrode). It was proved by Sir J. J. Thomson that these corpuscles—now known as *electrons*—carried negative electric charges and later it was established that they were the same in mass and electric charge whatever the gas used in the discharge tube and whatever the metal used as electrodes. Evidently electrons were constituents of matter, being torn from the atoms of the gas under the influence of the electrical forces prevailing. Further, electricity itself—whatever it may be—appeared to be itself atomic in character, the charge on the electron being now regarded as an indivisible unit of negative electricity. The mass of the electron was measured and found to be about $\frac{1}{1840}$ the mass of the hydrogen atom.

Experiments with the discharge tube revealed a further phenomenon; a stream of positively charged particles passed from the anode *A* to the cathode *C*. It was soon established that these particles were simply the atoms of the gas experimented with and that the positive charge carried by the atoms was equal in magnitude to, or might be twice or thrice the magnitude of the electronic charge. It appeared then that one or more electrons were torn from the individual atoms of the gas used in the discharge tube, leaving the latter charged with one or more units of positive electricity; in this condition the atoms are said to be *ionized.* The fact that the stream of ionized atoms can be deflected by electrical and magnetic forces was utilized to determine the mass of the atom of the gas concerned; the same principle, it may be added, was applied in determining the mass of the electron, the value of the electronic charge being found from other experiments. In particular the mass of the hydrogen atom was found to be such that about

600,000 million million million atoms

are required to make up one gram of the gas. Instead of dealing with the actual masses of the atoms it is more convenient to deal with their atomic weights which, it will be remembered, refer to the ratios of the atoms on the basis that the mass of the oxygen atom is denoted by 16.

A significant discovery was made by Sir J. J. Thomson in 1912. Experimenting with the inert gas neon in a discharge tube and applying electrical and magnetic forces to deflect the ionized atoms of the gas, he found that the atoms could be sorted out into two distinct streams, one consisting of atoms of atomic weight 20 and the other consisting of atoms of atomic weight 22. The atomic weight of neon as found by previous methods is 20·18; it now appeared that neon atoms are not all alike but are a mixture of two kinds, called *isotopes*, very nearly in the proportion of nine atoms of atomic weight 20 to one atom of atomic weight 22—these proportions give 20·20 as the atomic weight of the mixture.*

* There is also a minute proportion of an isotope of atomic weight 21.

A much more delicate sorting-apparatus, called the mass-spectrograph, was devised by Dr F. W. Aston in 1919. Examining the chemical elements one by one, Aston found that the great majority were mixtures of isotopes, the atomic weight of each isotope obeying, with very slight deviations, the whole-number rule. If we leave these deviations aside for the present,* it appears that Prout's hypothesis is re-established. Chlorine, for example, has the atomic weight 35·47 as obtained by chemical methods and, as we have seen, was one of the elements which helped to overthrow Prout's hypothesis; now it is known that chlorine consists of two isotopes, one of atomic weight 35 and the other of atomic weight 37, very nearly in the proportion of ten atoms of the former to three of the latter. Chemically, the two isotopes are indistinguishable, for they enter into chemical combinations in the same way; it is only by the refined physical processes alluded to that their separate existence can be detected. The element with the largest number of isotopes is tin (atomic weight 118·7) with ten, ranging from atoms of atomic weight 112 to atoms of atomic weight 124. Most elements, it is found, are mixtures of isotopes. Even hydrogen has a heavy isotope of atomic weight 2 to which the special name *deuterium* is given, and there is also evidence of the existence of a third isotope of atomic weight 3. It may be added that in Aston's experiments the relative abundance of the isotopes of an element can be determined; the atomic weight of the mixture, as found in Nature, can then be easily calculated and in all cases this is found to agree, within the limits of experimental error, with the atomic weight of the element derived by chemical methods.

The capacity of the discharge tube to provide surprises is not yet exhausted. In the early days of experiments with the discharge tube it was noticed that, under suitable conditions of pressure of the gas used, a bluish-green glow or fluorescence was produced on the glass walls of the tube; this effect was

* Despite their apparent insignificance, they are of immense importance in later developments.

associated with what we now know is the stream of electrons proceeding from the cathode. In 1895 Röntgen discovered that, when a metal disk was placed within the tube at an angle to the direction of the electron stream, highly penetrating rays were reflected out of the tube, one method of detection being based on their capacity for fogging a photographic plate, in the same way as light affects a plate. At first it was uncertain whether these 'rays' were a stream of corpuscles or a radiation with the general characteristics of light and so they were designated with the algebraic symbol of the unknown as X-rays. It is now known that X-rays are truly a radiation, their wave-lengths being exceedingly small when compared with the wave-lengths of light; it is this feature that accounts for their penetrating power. For example, X-rays can pass quite easily through the flesh of a hand but not so easily through the bones; consequently, a photograph of the bone structure can be readily obtained. It is not surprising that X-rays have proved of immense value in surgery and medicine. It was also found that X-rays falling on certain substances caused these to fluoresce; it was this connexion between X-rays and fluorescence that led to a revolutionary discovery in physics.

At the time of the discovery of X-rays Professor Henri Becquerel was interested in the phenomenon of phosphorescence, a peculiar property which some substances possess: if such a substance is exposed to sunlight, for example, and then removed to a dark room it will glow for some time afterwards in the same way as in the phenomenon of fluorescence already mentioned; evidently the substance has the power of storing the sunlight and then emitting it in distinctive colours at its leisure. Becquerel conceived the idea that possibly a phosphorescent substance might be capable of emitting X-rays —a reverse process to that already known, namely, the capacity of X-rays to cause fluorescence. He accordingly exposed various substances to sunlight, wrapped them up in black paper and laid each package on a photographic plate also protected by black paper; if any substance in this state emitted X-rays, which can easily penetrate several layers of black paper, this fact would be divulged by the fogging of the plate. However, all these experiments were unproductive of any positive result until he tried a salt of the rare metal uranium, the heaviest of all known elements, when the fogging effect was obtained.

But he made a further discovery of deeper significance unconnected with phosphorescence, for, if the salt were *not* exposed to sunlight or any other light, the photographic plate was still affected; moreover, the effect could be observed at any time, showing that the emission of some form of radiation was proceeding continuously; further, any salt of uranium possessed the same property, suggesting that it was the uranium itself that was responsible for the phenomenon. Here was a discovery of a new form of radiation, continuously and spontaneously emitted by uranium and capable of penetrating several layers of black paper.

Becquerel's discovery was published in 1896 and almost immediately Pierre and Marie Curie turned their attention to the investigation of the mysterious radiation. Their first problem was to devise a suitable apparatus for measuring the intensity of the radiation; using pure salts of uranium, they soon discovered that the intensity of the radiation was proportional to the actual amount of uranium used: twice the amount gave twice the intensity and so on. Further, the intensity was found to be independent of controllable extraneous conditions such as temperature. Marie Curie was soon convinced that the radiation was the result of some process—to which she later gave the name *radioactivity*—intimately connected with the uranium atoms. This atomic property might be conceivably shared by other elements and so, with this possibility in mind, Madame Curie undertook the laborious task of examining all the chemical elements; at first one only, namely thorium, was found to possess the radioactive properties exhibited by uranium.

Madame Curie now made a surprising discovery; she found that the intensity of the radiation emitted by a natural uranium ore called pitchblende, mined at Joachimsthal in Bohemia, was very many times greater than the intensity resulting from the known amount of uranium and thorium in the ore, as calculated from her previous investigations. Whatever doubt she had as to the accuracy of the experiments was dispelled by numerous repetitions. There could be only one explanation; the pitchblende must contain another and unknown radioactive element presumably in exceedingly minute quantities, for otherwise it could not have escaped the searching analysis of the chemists. At this stage Curie abandoned his own special researches to join his wife in what proved to be epoch-making investigations, the scientific and human partnership only ending with Curie's death in a Paris street-accident in 1906. For four years they toiled in a laboratory little better than a shed, under physical conditions that would have daunted all but the stoutest hearts, at the heroic task of extracting the unknown radioactive substance from vast quantities of pitchblende. They soon found that there were two sources of the unexplained radioactivity and one was quickly segregated; this was a new element polonium, so called after Madame Curie's native country, Poland. The second substance, which the Curies proposed to call *radium*, proved at first far more intractable but at length, in 1902, they succeeded in extracting a minute quantity of radium—a tenth of a gram—in the form of a chloride.

The discovery of radium was an outstanding landmark in the history of physical science and the Curies were rewarded with a half share of the Nobel Prize, the other half going to Becquerel. The radioactivity of radium proved to be more than a million times more intense than that of uranium. Curie, collaborating with medical men, showed that the radiation from radium destroyed diseased tissues; thus was begun the medical treatment known as radio-therapy which has been so familiar in our hospitals up to the present time.

The principal commercial supplies of radium are obtained from mineral ores at Joachimsthal (as already mentioned), near the Great Bear Lake in north-west Canada and at Katanga in the Belgian Congo. Radium is an extremely rare element and very difficult to isolate; it is not surprising that the commercial value of an ounce of radium is said to be in the neighbourhood of £150,000.

In 1898, a young New Zealand research student at the Cavendish Laboratory, Cambridge—Ernest Rutherford, later Lord Rutherford of Nelson—became keenly interested in Becquerel's strange discovery and the Curies' first researches. Soon he was embarked on the flood-tide of a remarkable series of discoveries, the earliest of which were made at McGill University, Montreal, where he was appointed Professor of Physics towards the end of 1898; between 1911 and 1919 his researches were carried on at Manchester University, and from 1919 till his death in 1937 at the Cavendish Laboratory. Rutherford first discovered that the penetrating radiation from uranium and thorium consists of two kinds, to which he gave the names *alpha*- and *beta*-rays, after the first two letters of the Greek alphabet; subsequently, a third kind of rays, designated *gamma*-rays, was discovered. The beta-rays were shown by Becquerel to be carriers of negative electricity and were actually streams of electrons shot out of the uranium and thorium atoms with speeds as great as 100,000 miles per second; the gamma-rays eventually proved to be a very penetrating radiation of the same character as X-rays but greatly superior to the latter in their capacity to pass through matter.

At first Rutherford was mainly preoccupied with the alpha-rays which were found to have a penetrating power only one-hundredth that of the beta-rays, being stopped by a thick sheet of paper or by 2 or 3 inches of air. The alpha-rays were soon proved to be positively charged material particles, each carrying two units of positive electricity (it will be remembered that the electron carries one unit of negative electricity) and each four times the weight of the hydrogen atom. As early as

1902 Rutherford and his collaborator Professor F. Soddy had suggested that the alpha particles were probably positively charged particles of helium, since helium was usually found imprisoned within the uranium and thorium ores, and in due course this surmise was definitely proved to be accurate. In the year just mentioned Rutherford and Soddy boldly stated that in the phenomena of radioactivity what is observed is nothing else than the spontaneous disintegration of the radioactive atoms and their metamorphosis into totally different atoms.

Hitherto, the atoms of the elements were imagined to be discrete entities constant in all physical and chemical characteristics, unchangeable and eternal; the dreams of the medieval alchemist as regards the transmutation of the elements had now come true, for in Nature such a transmutation was seen to occur and that without any adventitious aid. It could no longer be doubted that the atoms of the radioactive elements were unstable, ready to break up under the stimulus of some powerful internal atomic force. The study of the various disintegration products proved difficult, but the persistence and ingenuity of the atomic physicists were eventually rewarded. We shall consider the disintegration products of uranium.

Uranium is the heaviest element known in Nature, its atomic weight being about 238; actually, it is a mixture of three isotopes, of atomic weights 238, 235 and 234, the first predominates and is referred to as uranium I. Uranium I is the parent of a large family of elements, shown in Table VI,* of which radium and polonium are intermediate members. It will be seen that uranium I is transformed into uranium X_1, firing out an alpha particle in the process. The rate of breakup of the uranium I atoms is extremely slow: this is indicated in the last column of the table by what is called the 'half-value period' which is the time required by half of a given number of uranium I atoms to be transformed into uranium X_1 atoms;

* The principal details only are shown.

TABLE VI. *The Uranium Family*

	Atomic Weight	Particle or ray emitted	Half-value period
URANIUM I	238	α	4,500 million years
Uranium X_1	234	β	24 days
Uranium X_2	234	β	$1\frac{1}{5}$ minutes
Uranium II	234	α	340,000 years
Ionium	230	α	82,000 years
RADIUM	226	α, β, γ	1,600 years
Radon	222	α	$3\frac{3}{4}$ days
Radium A	218	α	3 minutes
Radium B	214	β, γ	27 minutes
Radium C	214	β, γ	20 minutes
Radium C′	214	α	One-millionth of a second
Radium D	210	β, γ	22 years
Radium E	210	β, γ	5 days
POLONIUM	210	α	140 days
LEAD	206	Inactive	—

in this particular transmutation, if we start with two million uranium I atoms, then after 4500 million years one million uranium I atoms will remain unchanged, one million uranium X_1 atoms will have been produced, and one million alpha particles will have been shot out.

The uranium X_1 atoms are, by comparison, short-lived for in 24 days half of them will be transformed into uranium X_2 atoms, the process involving the emission of a beta particle (that is, an electron) in each atomic transmutation. The half-value periods indicate the relative stability of the various members of the uranium family; the table shows that uranium I is by far the most stable and that radium C′ is by far the least stable, the rapidity of the disintegration of the latter being so great that of one million radium C′ atoms only five remain intact after one second. Points of interest in the table are that radium and polonium—the discoveries of the Curies—are products of uranium I in the chain of transmutations and that the final product is the inactive element lead; this last, for convenience designated uranium-lead, is a comparatively rare isotope, the ordinary lead usually found in deposits consisting of four isotopes of atomic weights 208, 207, 206 and 204 in order of abundance.

There are two other families of radioactive elements, namely, the thorium family and the actinium family, each characterized by similar disintegrations as in the uranium family and each ending up in an isotope of lead. It is to be remembered that the radioactive elements are the heaviest found in Nature and it was regarded at first as significant that their instability appeared to be associated with the presumed complexity of their atoms. However, in recent years three elements of moderate atomic weight—potassium, rubidium and samarium—have been shown to be mildly radioactive, the first two shooting out electrons and the third alpha particles; it is certain that such transmutations are properties of the rare isotopes of these elements.

In the hands of Lord Rutherford the alpha particles became the means of penetrating some of the closely guarded secrets of atomic structure. We remember that these particles are material, being in fact positively charged helium atoms—or, more particularly, helium nuclei as we shall see later—each four times the weight of a hydrogen atom and carrying two units of positive electric charge and ejected from the atom with speeds of the order of about 7000 miles per second. Hitherto, the belief that matter was atomic in character was the logical outcome of the study of innumerable experiments and phenomena in physics and chemistry but no individual atom had ever been identified. Now it was found that when alpha particles hit a screen coated with a layer of zinc sulphide, each alpha particle revealed its individuality by producing a scintillation, or minute explosion, on the screen; in this way the rate at which alpha particles are shot out of the radioactive elements was first measured. Other ingenious methods are now available for keeping track of the effects produced by individual alpha particles and similar charged particles.

One of Rutherford's most famous experiments was the bombardment of gold foil with the alpha particles shot out from radioactive elements, an experiment which with other established phenomena gave him the clue to atomic structure. In Fig. 14 we represent the main features of the experiment; a small amount of radium placed in a thin tube is the source of the alpha particles which, of course, are shot out from the radium in all directions; most of the alpha particles are stopped by the tube, but the remainder form a stream of projectiles directed along the length of the tube, just as bullets

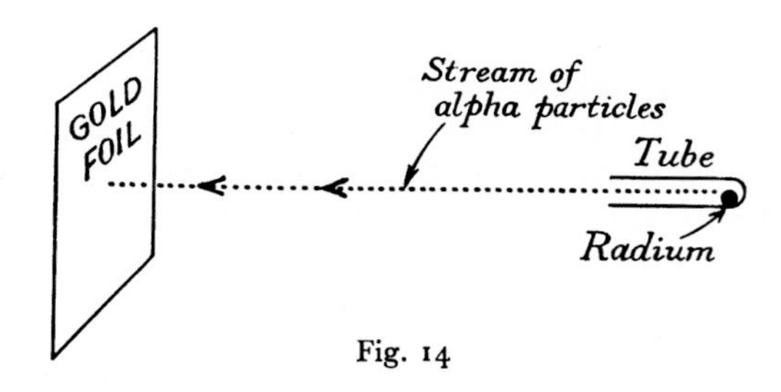

Fig. 14

are directed along the bore of a machine-gun. The foil against which the stream was directed was so thin that the majority of the alpha particles passed right through it; but some were thrown back from the foil at different angles and a few even returned almost on their outward tracks, scintillations on a zinc sulphide screen providing ocular evidence of the scattering phenomenon as it is called. It was just as wonderful —roughly in Lord Rutherford's words—as though a gunner fired a 15-inch gun at a sheet of paper and the 'cannon ball' came back to hit him. What could be the explanation of so extraordinary a phenomenon?

Rutherford's answer was to formulate his conception of the structure of the individual atom. An atom—say, of gold— consists of a small *nucleus* carrying a number of positive charges of electricity and containing practically the whole weight of the atom. As an atom under normal conditions is electrically neutral, which means that it is neither positively nor negatively charged, the positive charge on the nucleus must be balanced by negative charges elsewhere. These negative charges are provided by the appropriate number of electrons which are supposed to circulate in orbits around the nucleus in much the same way as planets circulate around the Sun—an atom being, as it is frequently expressed, a miniature solar system. This is not, however, a true analogy in all details, but for our purposes it may be regarded as providing a convenient picture. We have seen that the dimensions of the Solar System can be expressed in terms of the radius of the outermost planet Pluto; so in the same way the dimensions of the atom may be expressed in terms of the orbital radius of the outermost 'planetary' electron. We further recollect that the dimensions of the Sun and planets are very small compared with the dimensions of the Solar System and that, in fact, one of the most striking features of the Solar System is its 'emptiness'. So it is with the atom. The nucleus of the atom is, except for hydrogen, a complex structure as we shall see later, and the analogy with the Sun must not be pushed too far; however, it is found in the case of the gold atom, for example, that the nucleus is so small that, if we could imagine such nuclei to be laid in contact in a straight line, the number to the inch would be about 40 million millions; the radius of the atom, on the other hand, is roughly a thousand times that of the nucleus; it may be added that the radius of the electron is somewhat larger than the gold nucleus.

With this picture in mind one can now interpret the remarkable behaviour of the alpha particles fired at the gold foil in Rutherford's experiment. It must first be stated that an alpha particle is simply the *nucleus of a helium atom*, its dimensions being even smaller than those of electrons and gold nuclei. If then an alpha particle, carrying its two positive charges of electricity, passes fairly close to the positively charged nucleus of a gold atom, its path will be somewhat deviated from a straight line owing to the electric repulsion between the two particles involved; if the approach is extremely close, the alpha particle will be repelled by the nucleus in a direction almost opposite to that of its original course. Since the dimensions of the nucleus are so small compared with the dimensions of the atom, the great majority of the bombarding alpha particles will not come within effective range of a nucleus and will pass substantially undeviated through the empty atomic spaces and the interstices between neighbouring atoms.

This revolutionary conception of the atom to which Rutherford and Bohr were led in 1911 is a strange antithesis to the earlier conception of the atom as a single solid entity; like the Solar System, the atom in its spatial characteristics is mainly emptiness but, unlike the Solar System in which the law of gravitation is supreme, its properties are mainly derived from its electrical constitution.

The simplest atom is that of hydrogen, the lightest element; the nucleus consists of a particle carrying one positive charge of electricity and around it circulates a single electron, like the Moon circulating around the Earth. On account of its importance in atomic physics the hydrogen nucleus is designated by the special name of *proton*. Practically the whole mass of the atom is concentrated in the nucleus, the latter's mass being about 1840 times the mass of the electron. Since the nucleus carries one unit of positive electricity and the electron one unit of negative electricity, the complete atom is electrically neutral. This description applies to hydrogen under ordinary terrestrial conditions. If energy is supplied to a mass of hydrogen, say by heat, the planetary electron can take up a variety of orbital paths, thereby increasing the atomic dimensions, and if the treatment is sufficiently drastic, as in the discharge tube, the electrons are torn from their nuclei, giving rise to the electronic stream from the cathode to the anode and a stream of protons in the reverse direction. Protons derived by this or other means can be accelerated to great speeds by special devices and can be used as bombarding particles in the same way as alpha particles. Owing to its importance we may here indicate the dimensions of the proton as in the case of the gold nucleus; about 2500 million million protons go to make up an inch, if we suppose them to be laid in contact in a straight line. It may further be added that the dimensions of the electron are about 200 times those of the proton.

For many years it was believed that protons and electrons were the only elementary particles in Nature out of which matter was ultimately built up. Let us consider, on this view, the constitution of the helium atom—the second lightest element and next to hydrogen in simplicity of structure. We recollect that the atomic weight of helium is 4, that is, dis-

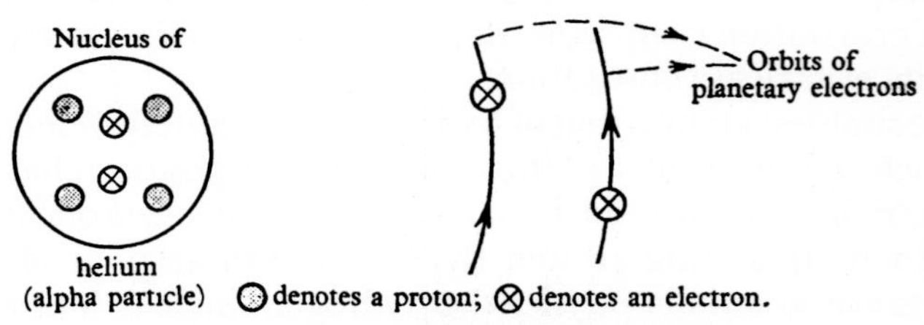

Fig. 15. Former model of the helium atom

regarding the masses of the planetary electrons, the mass of the helium nucleus is four times that of the hydrogen nucleus; if complex nuclei are built up of protons, the helium nucleus must then contain four protons. Further, as we have mentioned before, the nucleus as a whole carries two units of positive charge; consequently, as it consists of four protons each with a unit positive charge there must be two electrons, each with a unit negative charge, bound within the nucleus to give the correct net balance of two positive unit charges. Moreover, as the complete helium atom is electrically neutral the net charge of two positive units on the nucleus must be further balanced by two planetary electrons. The model of the helium atom was pictured very much as indicated, diagrammatically and without any pretence to correct scale-representation, in Fig. 15. Atoms of the other elements were supposed to be built up in a similar way. In later figures we represent a proton by *P* and an electron by *E*.

To take an example of a radioactive element, we consider uranium I (atomic weight 238); its nucleus was supposed to consist of 238 protons and 146 electrons, thus giving a net positive charge of 92 units; around the nucleus circulated 92 electrons to render the atom as a whole electrically neutral. The radioactive disintegration, which could only be a transformation of the nucleus, was then pictured as follows. The

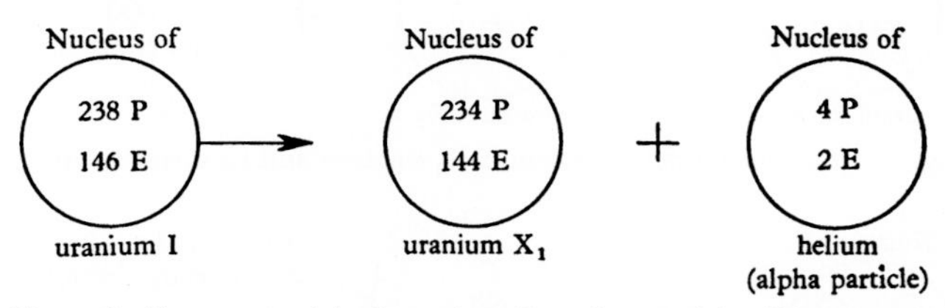

Fig. 16. Former model of uranium I nucleus and its disintegration

nucleus of the uranium I atom, being unstable, spontaneously disintegrates into two nuclei, one being uranium X_1 (see Table VI, p. 131) and the other an alpha particle (or helium nucleus); the latter is shot out with a speed of several thousands of miles per second, the energy necessary for expulsion being drawn from the store of energy imprisoned within the original nucleus. The disintegration of a nucleus of uranium I was represented as in Fig. 16 where the numbers of protons (*P*) and electrons (*E*) in the nuclei involved are shown. The nucleus of the uranium X_1 atom is also unstable disintegrating into uranium X_2 with the expulsion of a beta particle, that is, an electron. This was represented as in Fig. 17. The energy required for the expulsion of the beta particle is derived from the energy stored within the uranium X_1 nucleus. The energy of the gamma-rays, emitted in later transformations, is derived from similar sources.

But this conception of nuclear structure, however satisfactory it appeared to be at first, led later to insurmountable difficulties in the interpretation of other phenomena; these problems were resolved in a general way by the discovery of a new elementary particle, the *neutron*, by Sir James Chadwick in 1932. Its existence had been predicted by Lord Rutherford as far back as 1920. It is an *uncharged* particle with approximately the same mass as a proton. It was now

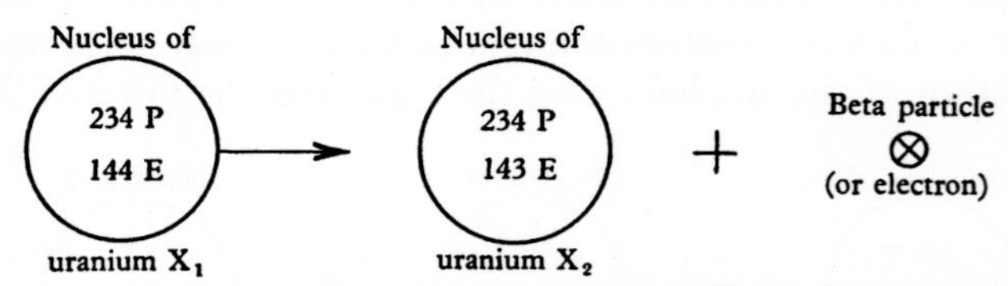

Fig. 17. Former model of uranium X_1 nucleus and its disintegration

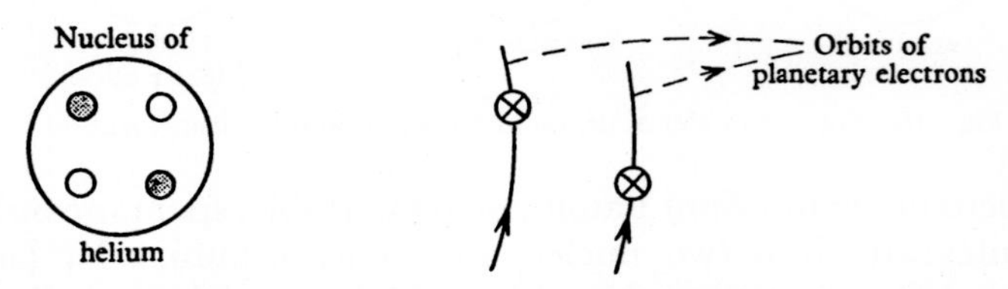

Fig. 18. Model of the helium atom

postulated that the nucleus of any atom, other than hydrogen, consisted only of protons and neutrons, the number of protons being sufficient to give the proper positive nuclear charge and the number of neutrons to give the balance of atomic weight.

The model of the helium nucleus (atomic weight 4, and positive charge 2) with its two planetary electrons is represented diagrammatically in Fig. 18. In later figures the neutron is designated by N and the proton and electron, as before, by P and E respectively.

In the same way the nucleus of the atom of uranium I consists of protons and neutrons, 92 of the former and 146 of the latter—the whole giving atomic weight 238 (the sum of 92 and 146); the positive charge is 92 and around the nucleus circulate a cloud of 92 electrons (not all shown in Fig. 19) thus rendering the complete atom electrically neutral.

The number of positive charges on the nucleus of a particular atom is an important constant of the atom, known as the *atomic number*. Alternatively, the atomic number is equivalent to the number of protons in the nucleus or the

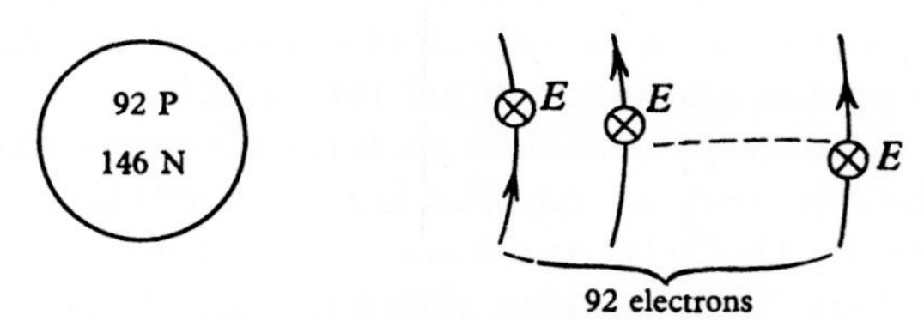

Fig. 19. Model of the uranium I atom

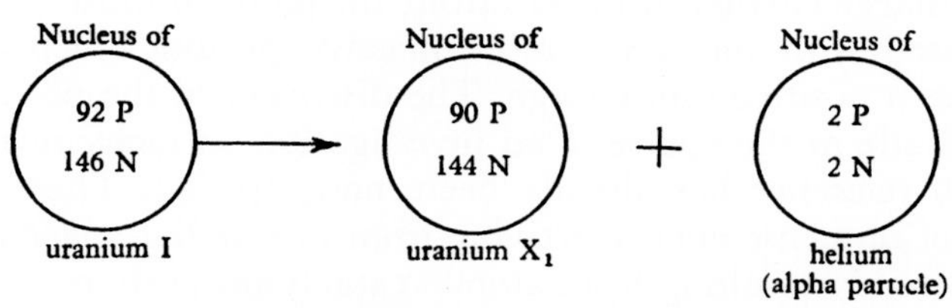

Fig. 20. Disintegration of the nucleus of uranium I

number of planetary electrons required to render the atom as a whole electrically neutral. The atomic numbers range from 1 (for hydrogen) to 92 (for uranium). The disintegration of the uranium I nucleus is represented in Fig. 20. So far this idea of the structure of nuclei seemed satisfactory. But what can be the mechanism of expelling a beta particle (or electron) from the nucleus of uranium X_1 in its transformation to uranium X_2, and in other radioactive disintegrations, when electrons are no longer supposed to be constituents of nuclei?

Hitherto we have given no indication that the three elementary particles—the proton, neutron and electron—are not the unchangeable and permanent entities supposed. But when we try to understand the bizarre microcosm of the atomic nucleus are we justified in making this supposition? The expulsion of beta particles from nuclei in radioactive disintegrations—for there can be little doubt that these particles originate in the nuclei—suggests a change of view and we ask if it is possible for a nuclear neutron to be capable of metamorphosis into a proton and electron, the latter being expelled as a beta particle by some sub-atomic process; we may even ask if the proton and electron are immutable.

Before we attempt to answer such questions we require to refer to the discovery in 1932 of a fourth elementary particle, the *positron*, by Dr C. D. Anderson of the California Institute of Technology. The positron resembles an electron except that it carries a positive unit charge instead of the negative unit charge carried by the electron; the positron must not be confused with the much more massive proton which also carries a positive unit charge. The discovery of the positron was made in the course of an investigation on *cosmic rays*, to which reference has already been made (p. 69). These are rays of immense energy arriving from interstellar space and presumably resulting from atomic cataclysms perhaps in the great extra-galactic nebulae. The impact of the cosmic rays on the gases of the upper terrestrial atmosphere produces streams of secondary rays of great penetrating power, so great that their effects can be detected at the bottom of deep lakes, and amongst the products of interaction between the cosmic rays and the atmospheric gases are positrons. Two facts emerged: first, the production of a positron was accompanied by the production of an electron, and second, the reverse process took place, that is, the recombination of a positron with any electron it encountered subsequent to its birth, resulting in the transformation of a pair of particles into radiation more penetrating than gamma-rays or X-rays.

The positron is a product of many radioactive disintegrations brought about artificially in the laboratory, as opposed to the natural radioactive transmutations so far considered.

The discovery in 1934 of *artificial radioactivity*, as it is now called, stands to the credit of Professor and Madame Joliot (the latter the daughter of Pierre and Marie Curie). Bombarding several of the lighter elements with alpha particles, they found that 'rays' continued to be emitted after the bombardment had ceased; this could only mean that the nucleus of the original element had been transformed into an unstable form which thereafter disintegrated spontaneously and ended as a stable element different from the original. The

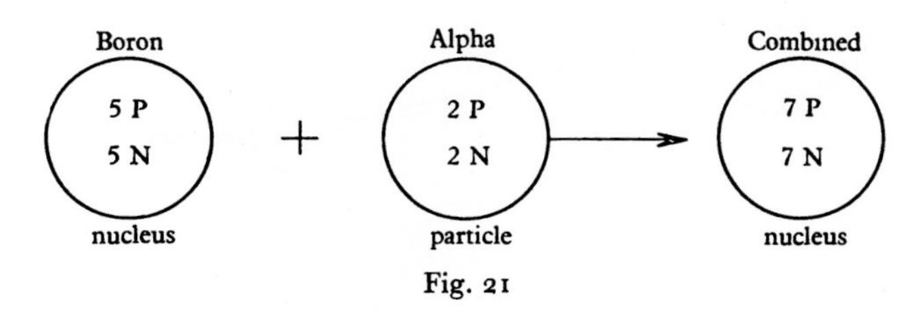

Fig. 21

process of bombardment in this connexion implies that a fast-moving alpha particle scores a direct hit on a nucleus of the original element and is captured by, and so to speak embedded within, the latter.

We take as illustration the bombardment of the rarer of the two isotopes of boron (atomic weight 10) with alpha particles. In the first stage the boron nucleus captures an alpha particle, absorbing both the material particle and its energy of motion; the boron nucleus consists of five protons and five neutrons and its capture of the alpha particle is represented in Fig. 21. The combined nucleus shown on the right is overcharged with energy and in the second stage it gets rid of surplus energy by shooting out a neutron, being reduced to an isotope of nitrogen (of atomic weight 13) otherwise unknown in Nature; this stage is represented in Fig. 22. Like the nuclei of uranium and radium this nucleus of the nitrogen isotope (13) is unstable and it disintegrates spontaneously after the fashion of the natural radioactive elements, emitting a positron (denoted by p) and ending up as a stable nucleus of the rarer isotope of carbon (of atomic weight 13) as represented in Fig. 23. The interpretation of this latter transmutation would seem to be that one proton of the nucleus of the nitrogen

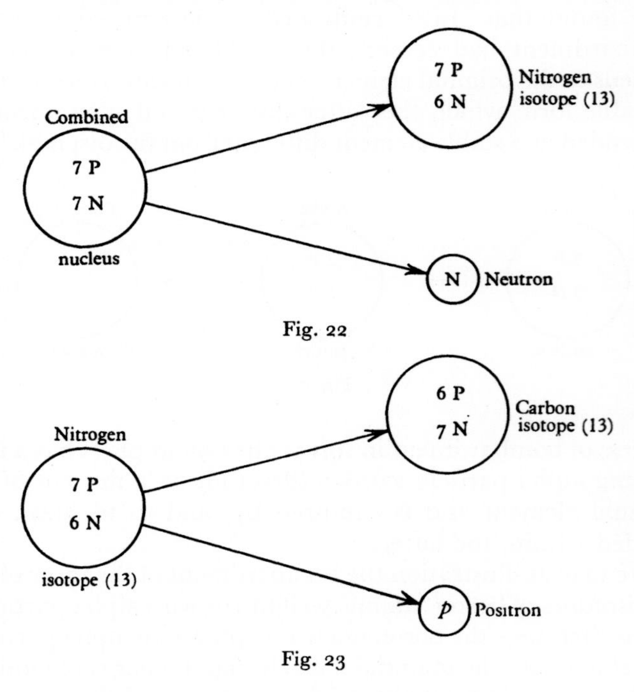

Fig. 22

Fig. 23

isotope is converted into one neutron and one positron, the latter being ejected; this transformation of the proton is illustrated in Fig. 24.

In the same way when aluminium (the atomic weight is 27 and it has no isotopes) is bombarded with alpha particles, an unstable isotope of phosphorus, otherwise unknown in Nature, is produced which decays radioactively with the emission of a positron, the end product being the rarest of the three isotopes of silicon.

But when aluminium is bombarded with neutrons a different transformation takes place. An aluminium nucleus captures a neutron and the nucleus so formed immediately gets rid of surplus energy by shooting out an alpha particle,

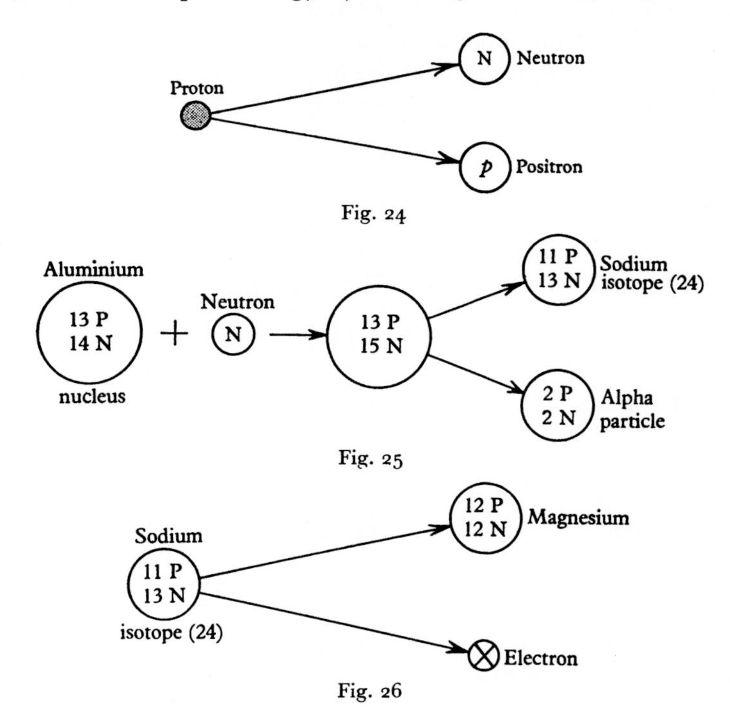

Fig. 24

Fig. 25

Fig. 26

being reduced in the process to a nucleus of an unstable isotope of sodium; we represent the capture of the neutron and the subsequent disintegration in Fig. 25. In the subsequent radioactive decay of the nucleus of the unstable sodium isotope, a beta particle—that is, an electron—is shot out and the end product is a nucleus of ordinary magnesium, the process being represented in Fig. 26. This last transformation indicates that one neutron in the nucleus of the sodium isotope is converted into a proton and an electron, the latter being ejected; the transformation of the neutron is illustrated in Fig. 27.

It appears then that the four fundamental particles which we have so far mentioned—proton, neutron, electron and positron—are not unchangeable and independent entities. These new conceptions make the nucleus and the constitution of matter more mysterious than ever.

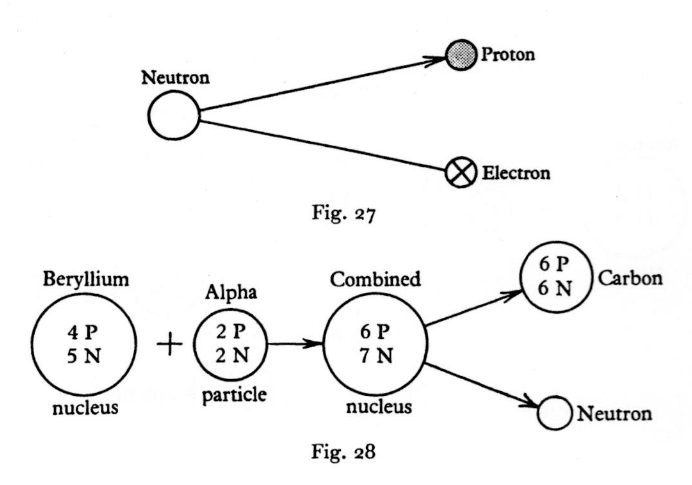

Fig. 27

Fig. 28

In introducing the neutron to the reader we merely stated the bald facts of its discovery. We now fill in the gaps in terms of the appropriate nuclear reaction. Chadwick bombarded the light element beryllium with alpha particles; neutrons were ejected and ordinary carbon was formed. The process is indicated in Fig. 28.

The Principle of Conservation of Energy

The reader may wonder why physicists are so confident that the nuclear reactions, illustrated above, actually represent what occurs during a nuclear bombardment by alpha particles and neutrons.* There is one aspect of atomic transmutations that has been only briefly mentioned so far, although it is of the utmost importance in elucidating the phenomena with which we have been dealing; we refer to the problem of energy. Consider first a disintegrating nucleus of uranium I. It shoots out an alpha particle with a velocity of several thousands of miles per second. The mass of the alpha particle is known and so the kinetic energy with which it leaves the uranium nucleus can be calculated. This kinetic energy can be produced only at the expense of the energy locked up within the nucleus; in fact Nature, in this instance, provides an automatic method of releasing energy. Again, in the bombardment of beryllium with alpha particles we apply energy in the form of the known kinetic energies of the alpha particles to transmute beryllium into carbon, with the emission of fast-moving neutrons, the kinetic energy of which can be measured. This reaction may, or may not, lead to a profitable exchange of kinetic energy even if every bombarding alpha particle hits a target.

All such reactions are governed by the great law of physics known as the law of conservation of energy which states, in the present connexion, that the sum-total of the energy (in whatever form this may appear) of the interacting particles just before transmutation takes place is precisely equal to the sum-total of the energy (again, in whatever form this may appear) of the particles and radiation produced as a result of the bombardment.† Previous to 1905 energy was a term restricted to kinetic energy, potential energy, heat and the energy of wave-motion exemplified in light and electromagnetic radiations, gamma-rays, X-rays and so on. There was a second law called the law of conservation of mass

* Protons and deuterons (the nuclei of the heavy hydrogen isotope of atomic weight 2) are also used as bombarding particles.

† In some reactions gamma-rays are emitted and the corresponding amount of radiant energy must be included in the final sum-total of energy produced.

which, in brief, stated that mass was indestructible so that, in a chemical action such as the formation of carbon dioxide from carbon and oxygen by ordinary combustion, the sum-total of the carbon and oxygen taking part was precisely equal to the mass of carbon dioxide formed. It is to be noticed that in this instance heat is produced, that is, energy; in other chemical actions heat, or some other form of energy, must be supplied to effect the combination of the interacting substances, as also for the decomposition of a chemical compound into its constituents. Einstein's 'special theory of relativity' of 1905 introduced the revolutionary idea that mass itself is a form of energy, the exact relationship between mass and its energy-equivalent being given by a simple formula.

As an example in which this new conception finds a place we consider the disintegration of a nucleus of uranium I into uranium X_1 with the expulsion of an alpha particle at a speed of about 7000 miles per second. If we restrict ourselves to the consideration of mass only, the process consists in the splitting of a nucleus of uranium I of atomic weight 238 into a nucleus of uranium X_1 of atomic weight 234 and a nucleus of helium of atomic weight 4; in other words the law of the conservation of mass appears to hold. But one feature of the disintegration that cannot be ignored is the production of the kinetic energy imparted to the alpha particle* and from the new point of view this kinetic energy is equivalent to so much mass. The disintegration process can then be represented by the following:

Mass of uranium I nucleus = Mass of uranium X_1 nucleus
+ Mass of alpha particle
+ Mass-equivalent of the kinetic energy carried off by the alpha particle.

We have mentioned earlier that Aston's very accurate deter-

* We neglect, for simplicity, the much smaller kinetic energy of recoil of the uranium X_1 nucleus.

mination, by the mass-spectrograph, of the atomic weights of the isotopes of the elements restored Prout's hypothesis of the whole-number rule of atomic weights (now, of course, applicable to the isotopes only) *except for minute differences*; it is these apparently negligible differences that are so important in the present connexion. The atomic weights of uranium I, uranium X_1 and an alpha particle are not precisely 238, 234 and 4 respectively; the atomic weight of uranium I is very slightly greater than the combined atomic weights of the two disintegration products and this excess amount is precisely the mass-equivalent of the kinetic energy of the alpha particle together with the recoil energy of the uranium X_1 nucleus (disregarding the restriction imposed in the footnote). To illustrate further: if the energy-equivalent of one pound of coal could be released at a suitable rate to drive a ship, with engines working at 50,000 horse-power, at 20 knots across the Atlantic and back, the energy supplied would be sufficient for about 250 crossings.

The old form of the law of conservation of mass has now to be abandoned and to-day physics recognizes only the law of conservation of energy—energy now being taken to include mass as well as kinetic energy, heat and the other forms already mentioned. This new conception of the law of conservation of energy and the accurate determination of atomic weights by the mass-spectrograph have been of fundamental importance in elucidating and interpreting the atomic processes involved in radioactivity, the bombardment of nuclei by fast-moving particles (alpha particles, protons, neutrons, etc.), and the interaction of energetic radiations, such as gamma-rays and cosmic rays, with matter. Some illustrations will be given in subsequent pages.

We can illustrate in some detail the principles of this new development in physics by considering the bombardment of lithium by protons (the nuclei of hydrogen atoms), carried out by Sir John Cockcroft and Dr E. T. S. Walton in 1932. Hitherto, the bombarding particles in such experiments had been the natural derivatives in radioactive transmutations. The protons used in the bombardment of lithium were, so to speak, manufactured articles, produced in the discharge-tube and accelerated under high voltages to very considerable velocities. It may be interpolated here that still greater velocities of protons and deuterons (the nuclei of heavy hydrogen) have been made possible by the invention of the *cyclotron*, an ingenious apparatus due to Dr E. O. Lawrence of the University of California, and by apparatus of later date producing still greater velocities.

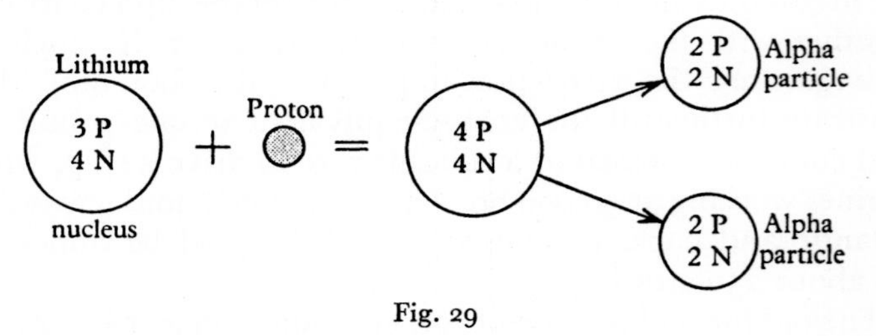

Fig. 29

The energy of the protons used as projectiles in Cockcroft and Walton's experiment was sufficient to transform a nucleus of lithium* into two alpha particles (Fig. 29). By the law of the conservation of energy, the total energy of the lithium nucleus (represented by its mass) and of the proton (represented by its mass and its kinetic energy) must equal the total energy of the two alpha particles (represented by their masses and kinetic energies). The delicate measurements made with the mass-spectrograph give the weight of the lithium nucleus and of the proton to be 7·0165 and 1·0076 respectively and of the helium nucleus to be 4·0028. Thus, considering the masses of the particles only, the successful bombardment of the lithium nucleus results in a diminution of mass from 8·0241 (i.e. 7·0165 + 1·0076) to 8·0056 (i.e. twice 4·0028);

* The third lightest element. Its nucleus consists of three protons and four neutrons.

consequently there is a diminution of mass amounting to 0·0185 or nearly 2% of the mass of a proton. This loss of mass is equivalent to the kinetic energy imparted to the two alpha particles (which actually fly apart in the experiment) less the kinetic energy of the bombarding proton of which the value is known. It transpires that the kinetic energy produced in the two alpha particles is about 150 times greater than the kinetic energy artificially applied to the bombarding proton; in other words, 150 times more energy is obtained than is expended, just as if in burning one ton of coal we miraculously got the equivalent, in heat-energy, of 150 tons. This was the first occasion on which subatomic energy was profitably tapped—in the sense that the energy extracted exceeded the energy applied. However, as only one proton is successful in hitting a target out of many million that are projected, the production of energy in this reaction is unprofitable in the commercial sense.

The Atomic Bomb

Before we end this somewhat brief account of the constitution of the atom and the phenomena of nuclear physics we take this opportunity of referring to the 'atomic bomb' which, in August 1945 at Hiroshima and Nagasaki in Japan, revolutionized warfare and at the same time provided modern civilization with an instrument capable of ending all organic life on our globe.* In 1939 Professor Otto Hahn of Berlin, a former research student of Rutherford at McGill University, discovered that slow-moving neutrons fired at the nuclei of the rare isotope of uranium of atomic weight 235 disrupted the nucleus into two unequal nuclei which flew apart with immense energy. We have seen that in the bombardment of

* President Truman's announcement of the unleashing of atomic energy on Hiroshima is historic: 'It is an atom bomb. It is the harnessing of the basic power of the Universe. The force [energy] from which the Sun draws its power has been loosed against those who brought war to the Far East.'

lithium with protons the energy produced is 150 times that applied; in the bombardment of uranium 235 the release of energy is on a far vaster scale; it is estimated that if all the nuclei in a pound of uranium 235 are disrupted the energy produced is roughly equal to the heat energy produced in the combustion of 1200 tons of coal.

But what makes the tapping of atomic energy possible on the gigantic scale of the atomic bomb is the fact that the disruption or fission of a uranium 235 nucleus is accompanied by the emission of several neutrons each of which is the potential destroyer of another uranium nucleus and the liberator of a further immense amount of energy. Here we have a chain action which, when once started, releases all the energy, capable of being tapped in this way, of all or most of the original quantity of uranium 235. Also, neutrons of high velocity can enter the nuclei of the ordinary uranium isotope of atomic weight 238 to form a trans-uranium element, named *plutonium*, which divides with an immense emission of energy in much the same way as in the fission of the nucleus of uranium 235; in both of these disintegrations the commonest and heaviest fragment appears to be the nucleus of the element barium of atomic weight 139.

All this, it need hardly be said, is only a brief recapitulation of the general principles involved in the fission of the uranium nuclei; the technical details relating to the atomic bomb remain a closely guarded secret. According to published accounts they involve in one form the use of heavy-water and, in another, graphite as a speed-reducer of the neutrons shot out of the nuclei. It may be surmised, however, that in the future the technique applied in the construction of the atomic bomb will be employed in tapping nuclear energy on a large scale for the benefit of civilization and not, as we must fervently pray, for its destruction; actually, small atomic 'piles', as they are called, have already been constructed for experimental purposes, the forerunners of the elaborate engineering plant that may revolutionize industry in our lifetime—provided that world peace becomes a reality and not merely a pious hope.

Many people deplore the rapidity of scientific advances in penetrating the mysteries of Nature and they go so far as to suggest that scientists should be subject to control, without specifying coherently how and by whom this control is to be exercised. Control of scientists means in fact the control of the inquiring mind and the frustration of those intellectual achievements that are, and should be, the glory of mankind. The atomic bomb, in a measure, crystallizes the discoveries in radioactivity and nuclear physics inaugurated by the simple experiment of placing some uranium sulphate on top of a photographic plate protected from light by layers of black paper; then came the discovery of radium (incidentally, with all its beneficent applications in medical and surgical treatment) by that heroic pair, Pierre and Marie Curie; then followed the superlative achievements of Rutherford and, somewhere in between, the revolutionary ideas of Einstein which gave a new orientation to atomic investigations and helped to make possible the immense progress in understanding the workings of Nature in the realm of the infinitely small. On whom should the control have fallen? On Becquerel, who made what at first sight appeared to be a harmless experiment? On the Curies, whose discovery of radium was applauded by the world, scientific and humanitarian? On Einstein, for adventuring too far in the realms of pure thought? On Rutherford and later experimental physicists, for following the instincts of the scientist? Should chemists and biologists and geologists and astronomers be controlled as well as physicists, for is it not a truism of the present day that all departments of science are interrelated? Where is the line to be drawn and who is to draw it?

Progress in scientific discovery has outrun progress in political philosophy, in ethics, in the promotion of better international relationships and in the propagation of religious principles; all these must catch up with science if civilization is to survive, and perhaps the power of destruction residing in the atomic bomb* will prove the medium whereby this stark truth may be driven home.

There is one further point of interest. If so much energy is locked up within the atom, where does this energy come from in the first instance? It is natural to suppose that the various ingredients of matter—protons, neutrons and electrons—are assembled and compressed, as it were, in some vast cosmic laboratory there to be transformed into the whole range of the ninety-two chemical elements. At one time it was thought that the immense temperatures prevailing in the deep interiors of the stars provided the requisite conditions for the fabrication of the complex nuclei. A process of this sort is, in fact, believed to take place in the Sun whereby our luminary draws on nuclear energy to maintain its expenditure of radiation, the process consisting of the synthesis of helium from hydrogen through the medium of the nuclei of carbon or nitrogen or both—this is known as the 'carbon-nitrogen cycle' with which we deal in the following chapter in greater detail. If, then, the carbon-nitrogen cycle is accepted as the effective nuclear reaction in the Sun, it would seem that carbon and nitrogen must have existed in the early days of the Sun's life when it was still a great globe of diffuse and relatively cold matter. Again, it is believed that the lithium-proton reaction of Cockcroft and Walton's experiment (p. 148) takes place under certain conditions; this consideration leads also to a similar conclusion as to the pre-stellar creation of the lithium nucleus. If this interpretation is correct, we arrive at the startling result that the atoms which we handle daily are older than the Sun and the stars.

* The latest development (1950) is the hydrogen bomb which, in one form, appears to depend for its destructive power on the release of energy by the bombardment of lithium by protons, as described on p. 148. The necessary temperature conditions for producing protons from hydrogen with sufficient speeds to penetrate the lithium nuclei are believed to be provided by the 'detonation' of a uranium bomb. The hydrogen bomb is thus a composite one.

THE RADIOACTIVITY METHOD OF MEASURING THE AGES OF THE ROCKS

In essence the principles of the method for determining the age of a rock containing a radioactive element are comparatively simple. We have seen earlier (p. 131) that uranium I disintegrates into uranium X_1 at such a rate that after 4500 million years half of the original amount of uranium I remains unchanged; also each atom on disintegration shoots out an alpha particle—or helium nucleus. This initial transmutation is followed by the series of changes recorded in Table VI on p. 131, the final solid product being the lead isotope of atomic weight 206, referred to conveniently as uranium-lead; in the whole series of changes from uranium I to uranium-lead eight alpha particles are shot out. As the rates of disintegration at all the stages are known it is possible to calculate the amount of uranium-lead and helium produced after, say, 1000 million years from a given amount of uranium I, and also the amount of uranium I still unchanged. In practice the calculation is inverted and the age of any rock containing uranium is deduced from the measured amount of uranium I still intact and the measured amount of uranium-lead (or of helium) found in the rock. The amounts of uranium I and lead are obtained, of course, from chemical analysis, and the amount of helium by a suitable physical process. The amount of a radioactive element in any sample of rock is extremely minute and one of the difficulties encountered in this kind of work is that associated with the measurement of the small quantities of the substances involved. Two methods of deriving the ages of the rocks are employed: first, the 'lead method' and second, the 'helium method'.

The 'lead method' as the name implies involves the measurement of the quantities of uranium I and uranium-lead in the rock. There are two principal complications. The first is that ordinary lead may also be present in the rock, thus adding to the difficulty of estimating the exact amount of uranium-lead genuinely produced in the radioactive transformations. The second complication is that thorium may also be present together with its final product, thorium-lead—the lead isotope of atomic weight 208. The age of the oldest rock found by this method is about 1700 million years.

The helium method, at first sight, does not seem to be a promising one. Helium is, of course, a gas and when it is formed in the heart of a rock it will strive to escape into the atmosphere. Some rocks are coarse-grained and are therefore unlikely to prevent the disappearance of helium from its rocky prison. On the other hand certain other rocks, such as basalt, are capable of retaining any helium formed during the radioactive processes, and it is these rocks that furnish the necessary data—namely, the amount of uranium still intact and the amount of helium contained—from which their ages can be calculated. Again there are complications if thorium or any other radioactive element is present. The success of the method depends on two assumptions: first, that the rock at the time of its solidification contained no helium and second, that when the helium was produced it was securely imprisoned within the rock. Much careful work was done in the pre-war years by both methods and the ages of the various rock-formations are now known with great precision (see Table V, p. 105).

From our point of view the salient result emerging from this close collaboration of physics and geology is that the age of the oldest rocks known is 1700 million years; possibly, still older rocks exist and will perhaps in due course reveal the secrets of their age. The Earth, as a planet, must be still older and it is not unreasonable to suggest that the solidification of its crust occurred not less than 2000 million years ago. Possibly the age of the Earth is as great as 3000 million years, or perhaps even more, but it seems to be unchallengeable that the age of our planet is at least 2000 million years.

The radioactive method has also been applied to determine the ages of meteorites—the chunks of cosmic matter arriving on the Earth from interplanetary or interstellar space. The ages, by which we mean as usual the intervals since solidification occurred, range from about 100 million years to nearly 3000 million years. We have already mentioned (p. 87) the bearing of these measurements on the general question as to the origin of meteorites; meanwhile it is sufficient to emphasize here the approximate agreement between the age of the Earth and of the oldest cosmic matter that can be directly handled.

In addition to specifying the ages of the rocks radioactivity also provided the explanation why Kelvin's conclusions, derived from his arguments relating to the cooling of the Earth, could no longer be accepted; it is only fair to remember, however, that Kelvin made the distinct qualification that his conclusions would be invalid only if sources of heat '*now unknown to us* are prepared in the great storehouse of creation'. As we have seen, energy is shot out of the radioactive elements in the form of alpha and beta particles and gamma-rays. The alpha particles are expelled from the nuclei of the elements, involved in the series of transmutations, with speeds ranging up to about 10,000 miles per second. In a rock the alpha particles are stopped almost immediately and the kinetic energy with which they are shot out is transformed into heat; the beta- and gamma-rays add their quota. The total rate of heat-production from radium was first measured by Pierre Curie in 1903, and from other radioactive elements by Rutherford a little later. Here, then, we have an internal source of heat in the Earth not envisaged in Kelvin's calculations; the Earth, in fact, must be very much older than Kelvin had predicted, as was first pointed out by Rutherford in a lecture at the Royal Institution, London, in 1904 on the energy emitted by radium. Rutherford's own account of this lecture is characteristic.* 'I came into the room, which was half dark, and presently spotted Lord Kelvin in the audience and realized that I was in for trouble at the last part of my speech dealing with the age of the Earth where my views

* *Rutherford*, by A. S. Eve, p. 107 (Cambridge, 1939).

conflicted with his. To my relief Kelvin fell fast asleep but, as I came to the important point, I saw the old boy sit up, open an eye and cock a baleful glance at me! Then a sudden inspiration came and I said that Lord Kelvin had limited the age of the Earth *provided no new source* [of heat] *was discovered.* "That prophetic utterance refers to what we are considering tonight, radium!" Behold! the old boy beamed upon me!'

Rutherford was the first, as just mentioned, to apply the radioactive method in estimating the age of the rocks; this was early in 1904. It is related * that, carrying a rod of pitchblende in his hand, he encountered his McGill colleague, the Professor of Geology, and inquired of him how old the Earth was supposed to be. When he was told that the estimate was about 100 million years, Rutherford quietly replied 'I *know* that this piece of pitchblende is 700 million years old'. As we have seen, Rutherford's assurance has been abundantly justified by later work.

In addition to specifying the minimum age of about 2000 million years for the Earth, the radioactive method can also be applied to assign an estimate for the maximum age of the crust, as was first done by Professor H. N. Russell of Princeton. Ordinary lead is mainly a mixture of three isotopes, of atomic weights 206, 207 and 208, occurring in varying proportions; the first is uranium-lead and the third is thorium-lead. The proportions of uranium, thorium and lead in the Earth's crust have been estimated. Suppose, first, that all the lead in the crust is uranium-lead; this must have been uranium originally and so, knowing the present amounts of uranium and of uranium-lead we can deduce the total amount of uranium when the Earth's crust was formed. The time required by the original amount of uranium to decay to its present amount can then be calculated. The result of the first calculation made on these lines indicated an age of 11,000 million years. In the same way if we assume that all the lead in the Earth's crust is thorium-lead, the age of the crust is about 8000

* *Rutherford*, p. 107.

million years. Since a proportion of ordinary lead is neither uranium-lead nor thorium-lead, the age of the crust is hardly likely to be as great as 8000 million years. Such calculations are necessarily somewhat rough since the estimate of the proportions of uranium, thorium and lead in the Earth's crust can hardly be expected to be precise; more recent calculations by Professor Russell indicate a maximum age of about 3000 million years. If the exact proportions of uranium, thorium, uranium-lead and thorium-lead were known, the age of the crust could then be derived with considerable precision. With the present evidence, however, we seem to be on safe ground in saying that the age of the Earth is at least 2000 million years and at most perhaps 4000 million years.

The substance of this chapter has shown that one important prop of Kelvin's time-scale for the age of the Earth has definitely gone. In the next chapter we shall see how the second effective prop—that relating to the time-scale for the Sun—has also suffered a similar fate.

CHAPTER VIII

THE ASTRONOMICAL EVIDENCE

In this chapter we consider three main lines of evidence which we may succinctly summarize as dealing with the age of the Sun, the age of the Moon and the age of the Universe. It is agreed that the age of the Earth is not likely to exceed that of the Sun; the Moon is not likely to be older than the Earth; the Earth's age, certainly, cannot exceed the age of the Universe. The sum-total of the astronomical evidence must then yield a pointer to the age of the Earth and the other planets which, it may be hoped—if we are to obtain a reasonably definite answer to our question—will not be inconsistent with the conclusions of the previous chapter.

Evidence from the Sun

We have seen that, according to Lord Kelvin, the Sun can only maintain its output of heat and light by drawing on potential energy and thereby contracting; further, the calculations showed that the Sun's age, reckoned from the time it was an immense globe of diffuse matter to the present time, could only be a score or two of millions of years. Even if it is supposed that the Sun can also call on energy derived from the known radioactive transmutations its age, although increased substantially, remains still less than the age of the oldest terrestrial rocks. Although on the contraction theory the rate at which the Sun's radius is diminishing can be calculated, this proves too minute to be detected by observation. The contraction theory must be checked in some other way if it is to remain a dominant argument in our problem; if it fails to be corroborated by observational tests, then we must look further afield for the main source of the Sun's radiant energy. The Sun is of course only one of a myriad of stars, and arguments relevant to the Sun might be expected to apply equally to the stars. It may not then seem strange to have to go to the stars to find indisputable proof that the contraction theory *by itself* is insufficient for the purpose in hand, although it must be recognized that, in general, the slow contraction of a star is a real phenomenon arising from the inexorable laws of evolution; one of the chief arguments will now be dealt with.

About a quarter of a century ago a delicate observational test was applied by Sir Arthur Eddington to a class of stars known as Cepheid variables—or simply *Cepheids*—so called from the bright star Delta Cephei in the constellation of Cepheus. Although, on the whole, the stars are constant in brightness there are many notable exceptions which fluctuate, some in a regular and predictable fashion and some in an irregular and unpredictable way; such stars are called variable stars, or simply *variables*. There are several types of variables, classified according to the manner in which the changes in brightness are brought about; the type with which we are concerned here is that of the Cepheids, the light-changes of which occur in a regular and characteristic way. The variability of Delta Cephei was discovered by John Goodricke, the heir to a Yorkshire baronetcy, in 1784 and since then the features of its light and other changes have been studied exhaustively. At its brightest—at *maximum*, as it is called—the star is easily visible to the naked eye; its brightness then diminishes for about $3\frac{2}{3}$ days when minimum brightness is reached; the star, still visible to the naked eye, is then about three-tenths as bright as it is at maximum; after minimum the brightness increases rather rapidly, the maximum being reached in about $1\frac{2}{3}$ days. The period between two successive maxima at the present time (1948) is known with very considerable exactitude—it is $5^{d}\ 8^{h}\ 47^{m}\ 26^{s}$. In this respect the star is thus like an accurate mechanism reproducing its cyclical changes with the precision of a reliable clock.

The periodical light-changes in Delta Cephei which we

have just mentioned are illustrated in Fig. 30; the curve is called a *light-curve* and in the present instance is based on photoelectric observations made by the author at Cambridge. The time is measured along *OA* and brightness along *OB*;

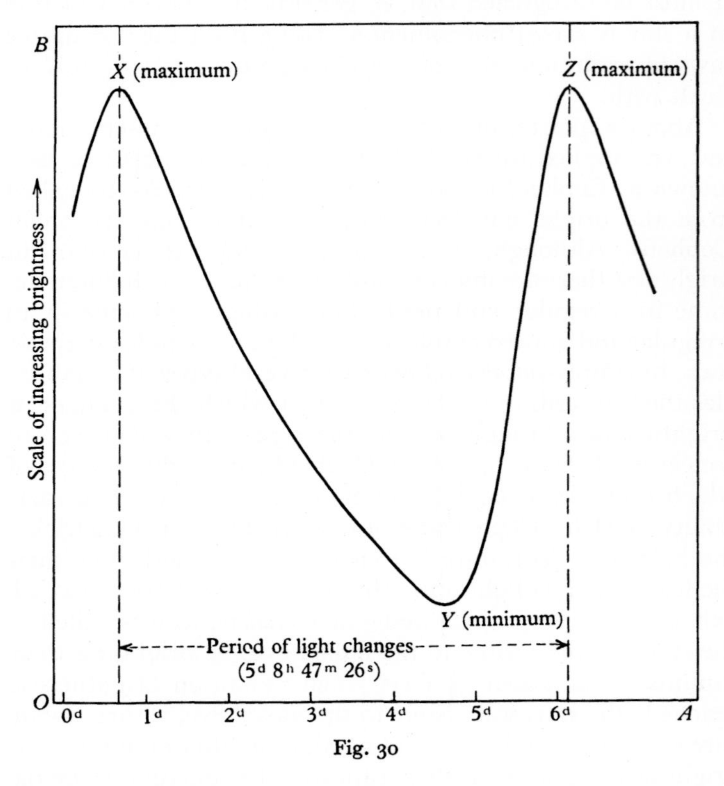

Fig. 30

each of the points *X* and *Z*, separated by the period of $5^d\ 8^h\ 47^m\ 26^s$, corresponds to maximum brightness (the maximum of the light-curve) while the point *Y* corresponds to minimum brightness; at maximum (*X* or *Z*) the star is about $3\frac{1}{3}$ times brighter than at minimum (*Y*).

In his published account of the discovery of the variability of Delta Cephei Goodricke remarked with astonishing penetration in one so young—he was then only twenty years of age, a deaf-mute from birth, and destined to die at the early age of twenty-two—that an inquiry of the kind he was describing 'may probably lead to some better knowledge of the fixed stars especially of their constitution and the cause of their remarkable changes'. In the following pages we shall describe in general terms how this remarkable prophecy of this brilliant youth came to fruition almost a century and a half later. The discovery of the periodicity of Delta Cephei was not Goodricke's only achievement. Two years earlier he had discovered the periodicity of the well-known variable star Algol, and in a famous paper to the Royal Society, of which he was elected a Fellow a fortnight before his untimely death, he ascribed the cause of light-variation to the eclipse of a bright star by another star revolving around it in much the same way as, on occasions, the Moon is responsible for partial eclipses of the Sun; this explanation was verified only after the lapse of 105 years from the date of Goodricke's discovery.

The variability of Algol and that of Beta Lyrae, discovered by Goodricke also in 1784, is a result of geometrical relationships and is thus of a wholly different character from the variability of Delta Cephei as we shall see immediately; such stars as Algol and Beta Lyrae are known as *eclipsing variables* and as such have little contribution to make to our general theme. Cepheids are found scattered fairly liberally about the Universe—in the immediate stellar environs of the Sun, in the globular clusters* and in the nearer spiral nebulae. The periods of light-changes range from a few hours up to about 100 days; in addition, other physical changes occur cyclically, but these do not concern us here.

In what follows we require to consider an important characteristic of a star, its *luminosity*. The luminosity of a star may first be most readily defined with reference to the Sun. The stars, of course, appear to have an enormous range in

* The variables in these objects are usually known as *cluster-variables*.

brightness, from Sirius (the brightest star in the sky) to the faintest star just detectable by the most powerful telescope. But this diversity in apparent brightness is, to a very large extent, due to the fact that the stars are at widely varying distances from us; other things being equal, the nearer a star is to us the brighter it will appear, just as the head-light of a motor-car a score of yards away appears very much brighter than when the car is a mile away. To be in a position to discuss the intrinsic brightnesses of the stars we ought to be able to compare their brightnesses if they were all placed at the *same* distance from us. The astronomer, of course, cannot move the stars about in the heavens; but if he knows the distance of a star he can easily calculate what its brightness would be if the star were situated at some selected standard distance from us; as he can also calculate how bright the Sun would be at this standard distance, it is then a simple matter to deduce how many times the star is brighter than the Sun, or vice versa. The ratio of the star's brightness to that of the Sun, obtained in this way, is called the star's *luminosity*. This definition involves the Sun as an intermediary standard of comparison. The rate at which the Sun is pouring out radiant energy (heat and light) can be measured; consequently, we have an alternative method of referring to a star's luminosity as the rate at which it radiates energy.

The range in luminosity is very great. There are stars so intrinsically bright that they are 10,000 times more luminous than the Sun; there are stars so intrinsically faint that the Sun is close on a million times more luminous than they are; the range, in fact, is far greater than that between a glow-worm and the most powerful searchlight. The feebly luminous stars are called *dwarfs* and the exceedingly luminous stars are called *giants*; the Sun itself, although not far from the dividing line, is classified as a dwarf.

One of Eddington's fundamental discoveries, applicable to the great majority of the stars, is the relationship between luminosity and mass—the more massive a star is, the greater is its luminosity; of course this mass-luminosity relationship is expressed in precise mathematical terms. At maximum Delta Cephei is almost 1000 times more luminous than the Sun; throughout its cycle of light-changes its average brightness is about 700 times that of the Sun. One or two other details concerning Delta Cephei are of interest; its average radius (its radius, we shall see, varies) is about 11 million miles (about twenty-five times the radius of the Sun); its mass is about $10\frac{1}{2}$ times that of the Sun; its average density is only about $\frac{1}{2500}$ of the Sun's average density which, it will be recollected, is about $1\frac{2}{5}$ times the density of water.

The most satisfactory theory accounting for the periodic physical changes exhibited by a Cepheid is the pulsation theory developed by Eddington; the star is supposed to expand and contract rhythmically, the period of a cycle of such changes being the period of light variation. How the oscillations or pulsations were originally started is a matter of surmise; however, this aspect of the problem hardly concerns us here. What emerges from the investigation is a significant relation between the period of pulsations and the star's average density—the smaller the density, the greater the period. Let us now follow out the implication of Kelvin's contraction theory in this connexion. If the star, on the average—that is, independent of the pulsations—contracts and keeps on contracting, its average density will continue to increase and, as a consequence of the relation between period and average density, it is deduced that the period of pulsations will decrease according to a precise mathematical formula. If Kelvin's contraction hypothesis is the only operative agency for supplying energy, it can be readily calculated that the average radius of Delta Cephei, smoothing out the periodic variations due to pulsation, must shrink at the present time by one part in 40,000 per annum—that is, by about 275 miles per annum; from this it follows that the star's average density increases annually by three parts in 40,000; from the relation between period and average density it is then easily calculated that the period must diminish at the rate of 17 seconds per annum. Accurate observations of the variability of Delta Cephei go back to 1840 and according to the best information the period then was $5^d\ 8^h\ 47^m\ 37^s$; the observations made by the author around 1933 with the Cambridge photoelectric photometer are consistent with a period of $5^d\ 8^h\ 47^m\ 28^s$, confirmed by other observers. Thus, in 93 years the period has diminished by 9 seconds, that is, by about one-tenth of a second per annum; from this it follows that, in a year, the radius has shrunk by 1 part in about 7 million, that is, at the rate of about $1\frac{1}{2}$ miles per annum—only a trifling rate as compared with the requirements of the contraction theory. There can be no question as to the inference to be drawn: the contraction theory can make no pretence to predict accurately the change in the period of light-variations and, consequently, must be judged to be unable by itself to account, except to a minute extent, for the maintenance of the star's output of radiant energy. The star must, accordingly, draw on some additional and much more abundant source of energy.

Before passing on to consider the source of stellar energy, we add some further remarks about Cepheids in general. A significant observational discovery was made about a third of a century ago by Miss Leavitt of Harvard Observatory; it was found that the periods of Cepheids * were related to their average luminosities in a definite way, the periods increasing with luminosity; this is called the period-luminosity relationship.† Without entering into technicalities we suppose that we can derive a curve expressing this relationship for Cepheids in general; accordingly, if we can determine the period of a Cepheid wherever it may be we can then deduce its average luminosity from the period-luminosity curve and, by a further step, its distance from us, as will be explained later. Now the

* Actually, the variables observed in the Lesser Magellanic Cloud.

† This relationship, established from observations, has so far been unaccounted for by the pulsation, or any other, theory.

period of light-variation of even a very faint Cepheid can be derived with comparative ease from a series of photographs the results of which are combined to establish a light-curve such as is illustrated in Fig. 30; the method consists in comparing the size, or the photographic density, of the images of the variable with the size, or density, of the images of one or more nearby stars of constant brightness; the brighter the variable in its cycle of light-changes the greater the size, or density, of its image on the plate. Having derived the period of variability we then obtain at once the star's average luminosity from the period-luminosity curve.

A further observational step enables us to deduce the distance of the variable from us; this step consists essentially in comparing the *apparent* average brightness of the variable during its cycle of light-changes with the apparent brightness of some particular star—which we denote by *A*—whose distance is known and whose luminosity is then calculable. To illustrate the method by taking a simple example, suppose that the apparent brightness of the star *A* is nine times the apparent average brightness of the variable.* Suppose further that the average luminosity of the variable as derived from the period-luminosity curve is the same as the known luminosity of *A*. In a sense we can suppose the two stars to be identical as regards luminosity but as *A appears* to be nine times brighter than the variable and as apparent brightness varies inversely as the square of the distance we conclude that the variable must be three times farther from us than *A*. When the circumstances are not as simple as we have supposed in the foregoing example, the calculation of distance is based essentially on the principles we have outlined. If the Cepheid observed is a member of a globular cluster (see Plate VI, facing p. 166) or of a spiral nebula such as the Great Nebula in Andromeda (Plate VII, facing p. 167), the distance of the

* The method of deriving this result is a photographic one, i.e. the comparison of the images of *A* and of the variable on plates taken under the same conditions.

cluster or of the nebula can be easily obtained. It is primarily by means of Cepheids that we are enabled to cast a sounding-line into the remote depths of space and obtain reliable measures of the distances of objects which at one time seemed far beyond our reach.

The Source of Stellar Energy

Since the contraction hypothesis fails to predict a sufficiently long age to the Sun and stars, we cannot ignore the challenge to discover a more successful solution. The clue in our search for the source of stellar energy can only be followed by understanding the physical conditions within a star. We consider first a giant star such as Betelgeuse or Antares—in either instance an immense globe of diffuse gas pouring out a stream of radiant energy across its surface into space. Since heat can only flow from a hotter region to a cooler, the temperature must increase from the surface of the star towards the centre. At the surface of Betelgeuse, for example, the temperature is about 3000° C.; the temperature at the centre must be much greater. Can we find with reasonable precision what the central temperature is? Again, the pressure of the stellar gas must also increase from the surface towards the centre, for the nearer to the centre the greater is the weight of gas to be supported by the pressure. Evidently the problem of ascertaining the physical constitution of a star is one of great complexity.

About 30 years ago, Sir Arthur Eddington began his famous investigations on the internal constitution of the stars which we shall attempt to describe very briefly. One novel feature in Eddington's researches concerned the pressure, at a point within the star, required to support the load of gas above. We are familiar with ordinary gas pressure, such as the pressure of air in a bicycle tyre, and this pressure depends on the density of the gas or air, if we assume that there is no variation in temperature. But with the enormous temperatures prevailing in the stellar interior there has to be added to the

Mt. Wilson Observatory

Globular Cluster in Hercules

PLATE VII

Warner and Swasey Observatory

Great Nebula in Andromeda

ordinary gas pressure the pressure exerted outwards by the out-flowing radiation—radiation pressure, as it is called—which gains rapidly in significance as higher temperatures are reached. We consider a small section such as *AB* in the interior of a star (Fig. 31). If the star is in equilibrium, that is, if we ignore the minute shrinkage for which the contraction theory is responsible, shown in the case of Delta Cephei, the

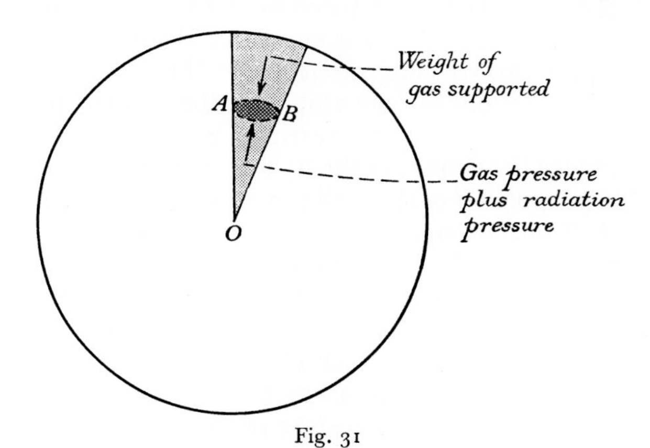

Fig. 31

gas pressure and the radiation pressure acting outwards across the section *AB* are together just capable of balancing the weight of gas above. Moreover, this gas acts as a screen through which the out-flowing radiation contrives slowly to pass, the stopping power of the gas being defined as its 'opacity'; the opacity is thus a new element in the physical background of the problem. To account for the maintenance of the out-flow of radiant energy from the central regions of a star towards the surface and thence into space, Eddington had to introduce the idea that energy was somehow or other produced in the deep interior at such a rate that when it leaked across the surface the output corresponded with the amount observed to be radiated by actual stars.

It might seem at first sight that precise knowledge of the

chemical constitution of the star would be necessary in any calculation; however, this apparent difficulty was easily resolved. As we have seen, the gas pressure in the interior of a star is a factor of prime importance. Further, in calculating this pressure one of the factors that matter is the average weight of the particles constituting the gas. Within the star, where very high temperatures prevail, the molecules of the chemical compounds are reduced to atoms and the atoms are stripped of most or all of their planetary electrons. The independent particles rushing about inside the star are now no longer molecules but mainly atomic nuclei and electrons and the average weight of these particles is found to be nearly equal to twice the weight of the hydrogen atom, unless a considerable proportion of the stellar material consists of hydrogen and helium. As an example consider the atom of oxygen: its weight is sixteen times that of the hydrogen atom and in its normal state it has eight planetary electrons. But if, within the star, the bonds holding the nucleus and the planetary electrons together are completely disrupted, the atom will now give rise to nine independent particles—one nucleus and eight electrons—and the average weight of these is practically $\frac{16}{9}$, or 1·8 approximately, times the weight of the hydrogen atom. In the case of iron (atomic weight 56), the complete atom consists of a nucleus and twenty-six electrons; if it is completely shattered, the twenty-seven independent particles have an average weight of $\frac{56}{27}$, or about 2·1, times the weight of the hydrogen atom. In these calculations the comparatively minute mass of the electrons has a negligible effect on the average atomic weight of the independent particles, since practically the entire mass of an atom is concentrated in its nucleus.

If we consider any other atom, the result is much the same; consequently, the chemical composition within the star is of little moment, for the average weight of nuclei and electrons derived from a mixture of the elements, with the exception of hydrogen and helium if these are abundant, is just a little greater than twice the weight of a hydrogen atom. With this simplification Eddington succeeded in calculating the luminosity of a giant star of given mass and this luminosity was found to be in satisfactory agreement with the observed luminosity; more generally he established what is called the mass-luminosity relationship which enabled the luminosity of any giant star to be deduced when its mass was known or, conversely, the mass to be calculated when the luminosity was known.

It is to be remarked that up to this point the theory, which involved the known laws of gases, was supposed to be concerned only with gaseous or giant stars such as Betelgeuse and Antares. But the surprising result emerged that the mass-luminosity relationship was successful in predicting the luminosity of comparatively dense stars such as the Sun, whose average density is $1\frac{2}{5}$ times that of water, and even of others with much greater densities. There could be only one conclusion: the material of these dense stars must be subject to the laws of gases and must in fact be in the gaseous state, despite appearances to the contrary. In a terrestrial gas such as air under ordinary conditions, the high degree of molecular mobility, which is the distinguishing feature of the gaseous state, depends on the sizes of the molecules and the number of molecules per unit volume. As we have seen, the independent particles within the star are nuclei and electrons compared with which the molecules of a terrestrial gas are exceedingly bulky entities; accordingly, a vastly greater number of the former can be packed into unit volume to produce a mobility comparable with that in an ordinary gas; thus, the density of stellar matter can be enormously high, perhaps as great as several thousand times the density of water, without its gaseous characteristics being imperilled. It would then seem that the same laws of gases are equally applicable to the diffuse giant stars and to dense stars such as the Sun; in other words, the mass-luminosity relationship has a theoretical basis, confirmed by observation, for *all* stars, with the exception, it must be added, of the special class of White Dwarfs for which the ordinary laws of gases are not applicable.*

It can be deduced from the theory that in the case of the Sun the central temperature is about 20,000,000° C. Moreover, it is found that the central temperature of the great majority of the stars is little different from that of the Sun. It would thus appear that the production of energy within a star must be closely related to this temperature; just as the water in the domestic kettle is transformed into steam at 100° C., so under suitable temperature conditions it would seem that energy boils off, as it were, from the central stellar material in a steady stream.

One thing we do know about stellar energy and that is the rate at which it is produced in the deep interior of the stars, for this must be equal to the known rate at which radiation leaves the star's surface on its journey into the void. There can be only one source of this energy, namely, the energy locked up in the atom or rather its nucleus. We have then to look for some process in which this nuclear or subatomic energy is being continuously tapped. The spontaneous disintegration of the radioactive elements is immediately suggested. It is deduced, however, that even if at one time the Sun, for example, consisted wholly of uranium the rate of supply of energy derived from the radioactive changes shown in Table VI (p. 131) is insufficient for the purpose in view. It may be argued, on the other hand, that in the stars there are more energetic trans-uranium radioactive elements not found on the Earth † which by their spontaneous transmutations supply energy at the requisite rate. But this is a speculation with which we can dispense if we can find a satisfactory solu-

* The best-known White Dwarf is the Companion of Sirius with a density about 60,000 times that of water.

† It will be recalled (p. 150) that in the uranium bomb, an unstable trans-uranium element of atomic number 94, called *plutonium*, is produced through the capture of neutrons by the nuclei of the uranium I atoms.

tion elsewhere. Instead of a star being an immense factory in which highly complex atoms break up radioactively into simpler atoms, may it not be that the exceptional conditions of temperature and pressure within a star are conducive to the building up of one or more complex atoms from the simplest of all, namely, hydrogen?

Discarding the radioactivity suggestion we are reduced to two possibilities for tapping the subatomic energy of stellar material; first, the release of energy in the synthesis of one or more elements from hydrogen and, second, the process to which we refer succinctly as the 'annihilation' of matter. In both, the relativity identification of mass with energy is invoked. Taking the 'annihilation' suggestion first, we note at once that there is an apparent contradiction, for energy and mass (as energy) are indestructible; what is implied is that the material of the star is converted, somehow or other, into radiant energy which eventually escapes into space; the mass of the star is thus gradually dissipated—in other words, the star continues to function as a luminous body only by shedding its mass.

If we limit the explanation to the case of electrons and protons which, at the immense temperatures prevailing within a star, are rushing about with high speeds, the 'annihilation' process is supposed to occur as follows: a collision of an electron with a proton, each carrying a unit electrical charge—the former, negative and the latter, positive—results in their destruction as material entities and the conversion of their mass into radiant energy, in the same way as electrons and positrons cancel out to yield a splash of radiation. Whatever the operative process, we can easily calculate the rate at which the Sun, for example, is converting its mass into radiation, for this rate must equal the known rate of the emission of light and heat energy across the Sun's surface into space; it is found that to maintain its present out-flow of radiant energy the Sun must be losing mass, by conversion into radiant energy, at the rate of about 4 million tons per second.

This may seem a prodigious rate but, in comparison with the Sun's mass of about

2000 million million million million tons,

the proportionate diminution of the Sun's mass is really proceeding at an extremely slow rate; even if we suppose that the present rate is maintained unchanged into the future nearly 16 million million years would be necessary for the complete disappearance of the Sun as a material body; actually, it is found that the rate diminishes as the Sun becomes less massive so that the duration of the future existence of the Sun, on the 'annihilation' hypothesis must be increased very substantially.

Looking back into the past we can calculate also how old the Sun must be if it began its life as a diffuse globe with a mass twice, thrice—and so on—its present mass; however massive it may have been, its age cannot exceed about 7 million million years—on the assumption, of course, that its mass is dissipated into radiation by the 'annihilation' process; this is usually known as the long time-scale.

One objection to the 'annihilation' of mass is that the mutual cancellation of electrons and protons is unknown in laboratory experiments; on the other hand, one known partner of the electron in a double suicide is the positron and this may represent the electron's potentiality for destruction under ordinary and stellar conditions. If we have to abandon the 'annihilation' hypothesis, we must attempt to discover an alternative mode for the generation of energy within the stars. There are other arguments, which we shall consider in due course, for believing that the long time-scale of several million million years is improbable and that a much shorter time-scale is to be preferred. If the truth lies in favour of the latter suggestion, the stars have only a comparatively short time in which to shed their mass by radiation, a process of evolution that is inevitable whatever the method of the release of energy in their interiors; but the proportionate diminution of mass is comparatively so small that the present-day mass of any star can be little different from its mass at birth.

We consider now the alternative hypothesis, the synthesis of one or more of the elements from hydrogen, as the possible mechanism for supplying energy in the interior of the Sun. We note first that, cosmically, hydrogen is a very abundant element and it is deduced by trustworthy arguments that about one-third of the solar mass at the present time consists of hydrogen; there can then be no question as to the availability of a sufficient stock of the basic 'fuel'. The next atom to hydrogen in simplicity of structure is helium and we consider the synthesis of helium out of hydrogen. We recollect that from the accurate measurements made with the mass-spectrograph the atomic weight of hydrogen on the usual scale is 1·0076 and the atomic weight of helium is 4·0028; we regard these numbers as applicable to the nuclei of these elements, for even in the complete atoms all but a negligible part of the mass resides in the nuclei. The synthesis of helium from hydrogen is then essentially the process of building up a helium nucleus from four protons in some way or other. If the synthesis can be achieved, a mass of 4·0304 (four times the mass of a proton) is converted into a mass of 4·0028 (the mass of a helium nucleus); in the process a mass of 0·0276 disappears and by the relativity doctrine of the equivalence of mass and energy this balance of 0·0276—roughly $2\frac{3}{4}\%$ of the mass of a proton—is converted into radiant energy. This is a much smaller transformation of mass into energy than in the 'annihilation' process, being only about $\frac{1}{147}$ of that occurring in the latter when the four protons are supposed to cancel out with four electrons.

Until recently the suggestion that the synthesis of helium out of hydrogen provided the flow of radiation within the Sun, or a star, was purely speculative, for it was not easy to understand how four protons could combine together to form a nucleus of helium consisting of two protons and two neutrons. In the previous chapter we had several instances of the artificial transmutation of nuclei (as distinct from the natural processes of radioactivity); might it not then be possible that in the deep interiors of the Sun and stars such a transmutation, or series of transmutations, would lead to the desired result? A condition of any suggested transformation must be that the rate of release of energy within the Sun, or star, must be equal to the known rate at which heat and light energy is emitted across the solar or stellar boundary; accordingly we have to 'try' various nuclear reactions and discard all those that do not fulfil this condition.

It was recently found independently, and almost simultaneously, by Dr Hans Bethe in the U.S.A. and Dr von Weizsäcker in Germany that one particular series of nuclear reactions could account satisfactorily for the present maintenance of the Sun's expenditure of radiant energy; this is known as the *carbon-nitrogen cycle* already mentioned on p. 152. We imagine that in the central regions of the Sun there is an abundance of protons, rushing about at the high speeds—about 300 miles per second on the average—appropriate to the temperature of 20,000,000° C., and also an adequate number of carbon or nitrogen nuclei. Let us assume first that carbon nuclei are sufficiently abundant. Sooner or later a proton will make a direct hit on a carbon nucleus, and on its capture—we can imagine that the proton becomes embedded in the carbon nucleus—a nucleus of an isotope of nitrogen (of atomic weight 13) is formed, gamma-rays being emitted in the process. This nitrogen isotope is unstable and, shooting out a positron, it becomes a nucleus of the isotope of carbon of atomic weight 13—this isotope is found in small proportions in ordinary carbon. This represents the end of the first stage in the transmutation, the conversion of the common carbon isotope, of atomic weight 12, into the rarer isotope of atomic weight 13. The process is indicated in the scheme (Fig. 32) in which the constitution of the nuclei in terms of the appropriate number of protons (P) and neutrons (N) is indicated, a positron being denoted by p and gamma-rays by γ.

In the second stage (Fig. 33) the nucleus of the carbon isotope of atomic weight 13 is hit in due course by a second proton; gamma-rays are emitted and the nucleus of ordinary nitrogen (atomic weight 14) is formed.

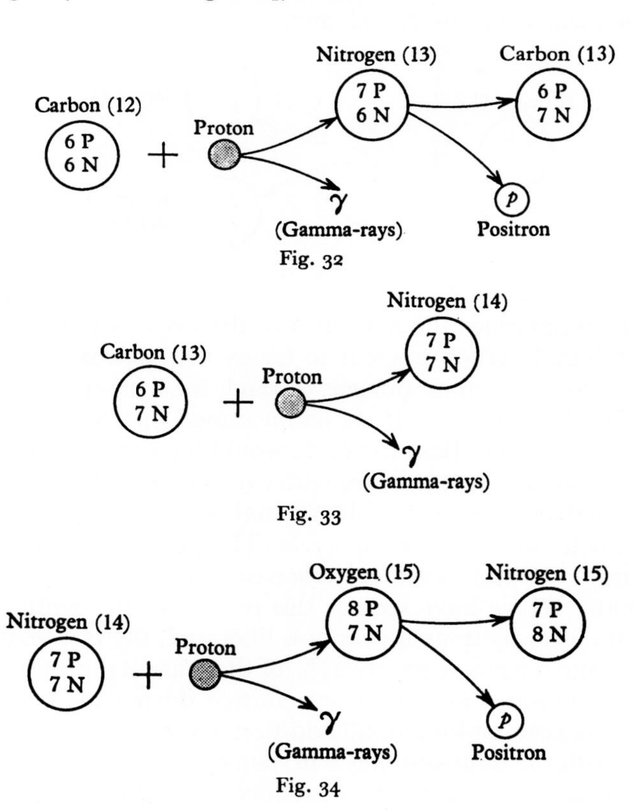

Fig. 32

Fig. 33

Fig. 34

In the third stage the nucleus of nitrogen (14) is hit by a third proton; gamma-rays are emitted and an unstable isotope of oxygen (atomic weight 15) is formed which decays into the stable isotope of nitrogen (atomic weight 15) with the emission of a positron, represented in Fig. 34.

In the final stage (Fig. 35) the stable nucleus of nitrogen (15) is hit by a fourth proton and the result is the formation of the *nucleus of the ordinary carbon isotope of atomic weight 12—identical to that with which we started*—together with a *nucleus of helium*, according to the scheme:

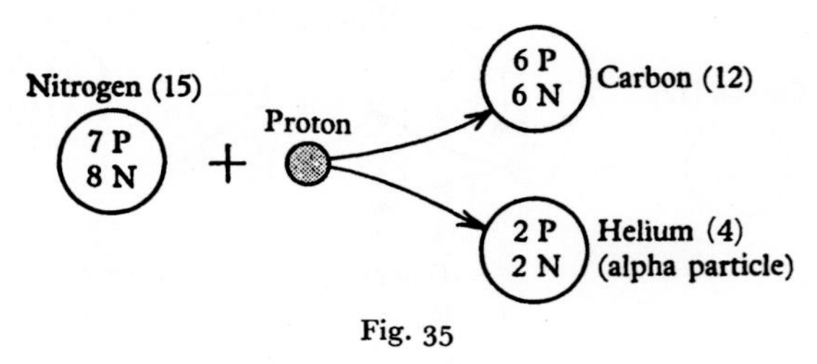

Fig. 35

To summarize: the net effect of the cycle is to reproduce the original carbon nucleus so far as structure is concerned and to form a helium nucleus through the agency of the four bombarding protons. If we had assumed that only nitrogen nuclei were abundant, the cycle would have started with the third stage, as shown above, with carbon nuclei formed in an intermediate stage and with the final restoration of the nitrogen nuclei to complete the cycle. The process then requires a suitable amount of either carbon or of nitrogen or a mixture of both. As we have seen in this process of the synthesis of helium from hydrogen, energy is liberated, the mass-equivalent being 0·0276 or nearly $2\frac{3}{4}\%$ of the mass of a proton; also the gamma-rays and positrons emitted during the course of the cycle account for a small addition to the energy liberated; as regards the positrons they will soon encounter free electrons and by mutual annihilation produce radiant energy.

It must be emphasized that the conditions for the success of the carbon-nitrogen cycle in accounting for the Sun's present rate of radiation is dependent on the high internal temperature of about 20,000,000° C., an adequate amount of carbon or of nitrogen, or of both, within the Sun and, of course, a plentiful supply of hydrogen; it is believed that these conditions are satisfied. The process will continue effectively until the store of hydrogen is substantially reduced or even exhausted; accordingly, the future existence of the Sun as a luminous body would appear to be guaranteed for thousands of millions of years.

But what can be said about the past history of our luminary in relation to the release of atomic energy? If, as is likely, the Sun condensed out of cold diffuse nebulous matter, the first process would undoubtedly be contraction, in the Kelvin sense, with an accompanying rise of temperature throughout. At some early stage the central temperature must be supposed to have reached a million degrees. This temperature is insufficiently high for the carbon-nitrogen cycle to be set in motion; but is it adequate for some other nuclear reaction whereby atomic energy can be tapped? It is found that at temperatures of the order of that just mentioned nuclear reactions involving the bombardment of the nuclei of the lighter elements—such as lithium, beryllium and boron—with protons releases energy in significant quantities.

A typical reaction of the kind envisaged is that of the bombardment of lithium (atomic weight 7) with protons, helium being formed; this, in fact, is the reaction artificially produced in Cockcroft and Walton's experiments described earlier (p. 148). In several other reactions involving the lighter elements helium is also produced. It would thus appear that in the Sun's comparative youth the lighter elements, including hydrogen but not helium, provide the 'fuel' for maintaining the Sun's luminosity—not necessarily the same as at present—over a considerable part of its life and this stage is effectively terminated only when the supplies of lithium, etc., are substantially exhausted. The process accounts for the fact that lithium, beryllium and boron occur only in minute proportions in the solar atmosphere; as the relative abundance of these light elements in terrestrial rocks is also extremely meagre it would seem to follow that, if the Earth were originally derived from the Sun, its birth must have occurred when the exhaustion of the lighter elements in the Sun—assuming these to have been originally in moderate supply—had proceeded a considerable way.

It may be recollected that the lithium-proton reaction is a specially energetic one. Although at temperatures of a few hundred thousand degrees the release of energy occurs at a comparatively moderate rate, at much higher temperatures the rate is immensely increased, becoming explosive in character. Perhaps the eruptive prominences so frequently observed on the Sun are the results of nuclear reactions at some depth below the solar surface where the temperature is sufficiently high and where a moderate concentration of such light elements as lithium occurs through fortuitous circumstances.

After the effective exhaustion of the lighter elements the Sun is unable at first to draw on further stores of subatomic energy; it must then contract, becoming hotter in the process, and only when the central temperature approaches 20,000,000° C. are the conditions suitable for the operation of the carbon-nitrogen cycle; this, as we have seen, is the process which is believed to occur at present and which will ensure a long future to the Sun as a luminous body.

This brief description* of the principal stages in the Sun's evolutionary history based on nuclear bombardments with protons does not include any estimate as to the Sun's age, for the original chemical composition of the Sun is unknown. But with reasonable assumptions we can make an approximate guess. Suppose that hydrogen and lithium in the Sun originally amounted to 40 and 5% respectively of the Sun's mass. For simplicity we confine the discussion to the lithium-proton reaction and the carbon-nitrogen cycle. Consider the former. We have seen (p. 148) that in a reaction involving one lithium nucleus and one proton nearly 2% of the mass of a proton is converted into energy which finally escapes into space as heat and light energy. The Sun's mass will conse-

* The same, or similar, arguments apply in general to the stars.

quently be diminished by a small but calculable proportion and it is found that the time required for the transmutation of lithium and hydrogen into helium is about 2000 million years; the amount of hydrogen used up in the process is about five-sevenths of 1% of the Sun's mass.

After the completion of the lithium reaction the Sun will contract, as in Kelvin's hypothesis, until the internal temperature is sufficient to stimulate the carbon-nitrogen cycle—the interval required for the necessary contraction can be only a few million years, which is negligible in comparison with the much longer interval just derived for the operation of the lithium reaction. Assuming that the proportion of hydrogen in the Sun at present is one-third of the Sun's mass, there remains a mass of hydrogen of about 6% of the Sun's mass available for the carbon-nitrogen cycle; in this process, as we have seen (p. 174), four protons are converted effectively into one nucleus of helium with a diminution of mass nearly $2\frac{3}{4}$% of the mass of a proton. As before, the resulting diminution of the Sun's mass can be calculated and the time required for the operation of the cycle up to the present time is about 6000 million years. The total time, then, required for the operation of the two reactions is about 8000 million years; on the hypotheses stated this represents the age of the Sun since its initial state as a diffuse globe.

We can submit our estimate to scrutiny by considering the amount of helium produced in the course of the two processes; this amount is close to 12% of the Sun's mass. The percentage of helium at present in the Sun must be greater, unless we assume that in its original state the Sun contained no helium at all—a most unlikely suggestion. If we allow a modest 3% of helium in the Sun's early youth, the present amount of helium is then 15%, on the hypotheses made. What the actual helium content is at present is not known with precision, but it would seem that our estimate of 15% is excessive. The Sun's age must then be much less than 8000 million years.

It was at one time considered likely that the Sun, with its

immense central temperature and pressure, might be a vast natural laboratory in which the heavier elements were synthesized out of the lighter. But it would seem—unless we throw overboard the arguments based on the lithium-proton reaction and the carbon-nitrogen cycle—that the heavier elements must have existed before the Sun started on its course as a luminous body. Where and when were the complex nuclei forged? We have to answer that we do not know.

Expansion of the Universe

The stars which we see in the sky on any clear night form part of a vast stellar system called the *Galaxy* or the *Galactic System*. Despite the fact that the Sun and its attendant planets are immersed within this immense aggregation of stars, several lines of investigation have enabled us to form a fairly clear and accurate picture of the shape and extent of the Galaxy, at any rate so far as the great majority of its stellar inhabitants are concerned. The main conclusion is that the system is shaped like a lens or bun (represented diagrammatically in Fig. 36) and within this space are enclosed about 50,000 million stars, together with interstellar clouds of gas and particles rivalling the stellar population in mass; the principal diameter, *AB*, of the lens or bun is about 100,000 light-years* in extent, and the 'thickness' of the system (*CD* in the figure) is perhaps 20,000 light-years.

Outside the lens or bun and forming a roughly spherical system are the great globular clusters—about a hundred in all, so far as is known—each a concentration of thousands of stars. Plate VI (facing p. 166), shows one of the best-known of these clusters. The Sun is situated quite close to the principal central plane of the lens (this plane is shown shaded in Fig. 36) and at a distance of about 30,000 light-years from the centre, *O*, of the system; a cross in Fig. 36 indicates the Sun's position. Outside the main concentration of stars in

* A light-year is about 6 million million miles.

the lens-shaped space, there is a scattered distribution of stars, comparatively few in number, giving the system a greater extent than is indicated by the lens-shaped configuration.

The whole system is rotating about the diameter *COD*—this is the phenomenon known as *galactic rotation*—in the sense

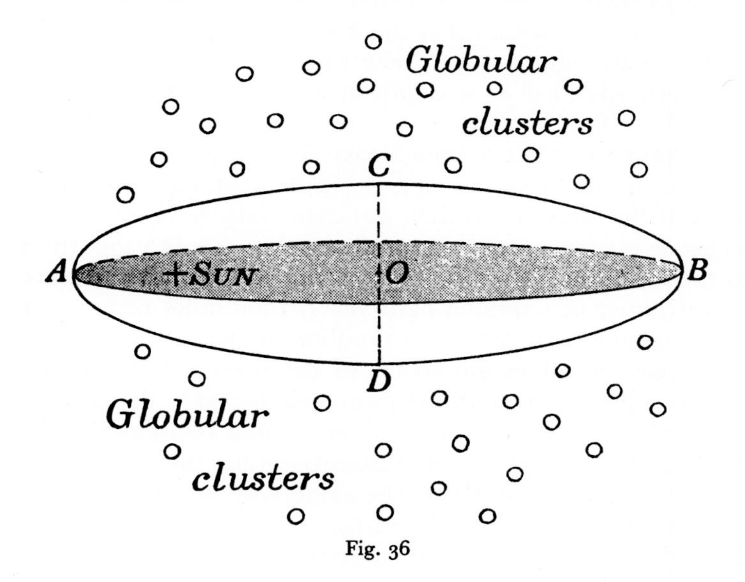

Fig. 36

that the stars describe vast orbits around the galactic centre *O* in much the same way as the planets describe orbits around the Sun; for example, the stars in the immediate neighbourhood of the Sun have orbital periods close to 300 million years. Such in brief are the main features of the Galactic System to which the Sun and its family of planets belong.

Looking out into space beyond the confines of our Galactic System we see immense numbers of objects variously described as 'spiral nebulae', 'extra-galactic nebulae', 'island universes' or 'external galaxies', each of which is believed to be a replica—in some respects, at least—of our own Galaxy and on much the same scale. Plate VII (facing p. 167), shows one of the nearest of these objects, the Great Nebula in Andromeda.

In some of the nearest of the spiral nebulae Cepheid variables have been discovered and their distances have been deduced by the method indicated on p. 165. Further criteria for deriving the distances of the fainter and, presumably, the more distant objects have been utilized and, so far, space has been sounded to a depth of about 500 million light-years.

Perhaps the most extraordinary feature about the spiral nebulae is that they all appear to be rushing away from us, as if the Earth or, rather, our own Galactic System were a cosmic plague-spot, with speeds that increase in proportion to their distances from us; 100 miles per second for nebulae at a distance of a million light-years, 1000 miles per second for nebulae at a distance of 10 million light-years and so on, the largest speed measured up to the present being about 25,000 miles per second for faint nebulae at a distance of 250 million light-years. This observational relationship between speed of recession and distance implies that the system of nebulae is expanding in the same way as the distances between marks made on the surface of a balloon increase as the balloon is blown to greater dimensions; what then appears at first to be a unanimous aversion of the nebulae to our society is now inferred to be a general phenomenon—whatever the nebula, all the others seem to be receding from it. The speed of recession for a particular nebula is obtained from its spectrum. It is known that, if a body is receding from us, the absorption lines—for example, those relating to the H and K lines of calcium which mainly supply the information as to nebular speeds—are displaced towards the red end of the spectrum by an amount proportional to the speed; in other words their wave-lengths appear longer than they would be if the object responsible were stationary with reference to the Earth. As the change in wave-length, called the *red-shift*, can be measured, the speed of recession can be deduced.* Although various attempts have been made to account for the red-shift in the nebular spectra on grounds other than velocity effects, all such efforts have failed and now there seems to be little doubt that the message of the displaced absorption lines is that the system of Galaxies is expanding in the way already indicated. So much for observation.

To turn now to theory. The theory of relativity asserts that space is expanding according to the same law which governs the speed of recession and distance of the nebulae. It is true that relativity predicts *either* an expansion *or* a contraction, but the observed phenomena associated with the distant Galaxies lead us to infer that the second alternative should be ruled out; and the interpretation of the nebular speeds is simply that space in expanding carries the nebulae, so to speak, with it. Although relativity theory by itself is unable to predict the numerical value of the rate of expansion, Sir Arthur Eddington succeeded in calculating what this rate should be, in a recondite investigation combining both relativity theory and the quantum theory of the atom; his result is in substantial agreement with the observed rate, after allowing for factors which need not concern us here. Although Eddington's arguments have not won anything like general acceptance, since there are few who are competent to pronounce judgement, it does seem an extraordinary achievement to calculate almost by pure theory what is happening in the farthest depths of space.

From our point of view the significant feature about the rate of expansion of the Universe is its magnitude; it is calculated that the system of Galaxies has doubled its dimensions

* For example, the H line of calcium has a normal wave-length of 3968·6 Ångström units (one Ångström unit—denoted by Å.—is, it will be recollected, one hundred-millionth part of a centimetre); the wave-length of this line measured in the spectrum of the most distant nebula for which information is available is about 4274 A., from which the speed of recession is calculated to be 25,000 miles per second.

in 1300 million years. If we go back 2000 or 3000 million years, the Universe must have been still more contracted and, clearly, there must be some limit of contraction beyond which it is impossible to go. It would then seem that the long time-scale of 7 million million years, considered previously, must be entirely ruled out of court and that we are left with a time-scale of a few thousand million years within which the Universe has reached its present development from an initial state of high concentration. Beyond this, in time, is mystery. Is it in this primeval stage that the nuclei of the elements were built up, later to be gathered into vast aggregations of matter forming the initial states of the great Galaxies out of which smaller condensations—the stars—are born? Whatever our speculations into the dim past of Creation, it would seem that the phenomenon of the expansion of the Universe puts a limit of a few thousand million years to the ages of the Sun and stars, the Earth and planets.

Evidence from the Moon

In an earlier chapter (p. 113) we discussed the effect of tidal friction and saw how, due to this cause, the day was lengthening slowly (at present, at the rate of 1 second in about 120,000 years); a secondary result is that the Moon is slowly receding from the Earth at present, at the rate of about 5 feet per century. Going back in time we infer that the Earth was rotating much more rapidly and that the Moon was much nearer the Earth. Further, the closer the Moon was to the Earth, the greater must have been the tidal effect produced by the Moon both in the shallow seas and in the broad oceans. It is not to be supposed, however, that the actual distribution of the seas and oceans was the same a thousand million years ago as it is at present; but, assuming that tidal friction in any past epoch was operative in much the same way and at much the same rate as at present, it is possible to calculate the Moon's distance from the Earth at any time in the past. In particular, the researches of Sir George Darwin and Professor

Harold Jeffreys have suggested that about 4000 million years ago the Moon's centre was but 8000 miles from the Earth's centre; then the Earth was rotating very fast in a period of about 4 hours*—and the Moon was revolving about the Earth also in a period of 4 hours. The further suggestion seems to be unavoidable that the Moon had then just been formed from the Earth, from which it follows that the age of the Moon is about 4000 million years at most.

We may take this opportunity of sketching what, at one time, appeared to be the most plausible theory to account for the formation of the Moon, known as the 'resonance theory'. As we have seen, the Earth's rotational period when the Moon was very close to the Earth was about 4 hours and this would still be, very nearly, the rotational period before the Moon broke away from the Earth as a separate body. It is to be supposed that at this point the Earth—or rather the composite body—was fluid, a state attained, presumably, after a comparatively short interval from its formation, and that tides were produced as a result of the Sun's attraction, high tide following low tide at any point on the Earth in an interval of about two hours. But this is also the period of disturbance for a body of the size and mass of the Earth and, accordingly, what is known as a 'resonance effect' is produced, one result of which would be the continuous increase in the height of the tides; thus the fluid body would become very much elongated, almost like a rather short cigar in shape, and when the elongation was sufficiently large, the stability of this body would be seriously jeopardized with the inevitable consequence that the mass would eventually break up into two parts, one identified with the Earth and the other with the Moon. Owing to the mutual tidal action between the Earth and the Moon, at this stage both fluid, the separation of the two bodies would proceed rapidly. In due course crusts would form as a result of cooling and when the terrestrial oceans appeared the mechanism of tidal friction

* According to our present time-reckoning.

would operate in the sense already described, bringing the Earth-Moon system to its present state. In recent years this theory of the Moon's formation has encountered several difficulties; but there can be little doubt that, whatever the process by which the Moon came into being as an independent body, the ages of the Moon and the Earth are almost identical and that the estimate of 4000 million years for the close approach of the two bodies, as derived from the arguments of tidal friction, must be substantially the age concerned.

The future course of the Earth-Moon system can also be traced; tidal friction will continue, of course, to slow up the Earth's rotation and also to increase the separation of the two bodies, thereby increasing the Moon's orbital period. It is found that after about 50,000 million years the Earth's rotational period and the Moon's orbital period will both be about 47 days, the Moon's average distance from the Earth being then about 350,000 miles (the average distance at present is about 240,000 miles). The equality of the two periods just mentioned means that the Earth will be turning the same face towards the Moon and consequently the lunar tides will temporarily cease to have any effect on the Earth's rotation.

The solar tides, however, will continue to play their part in slowing up the Earth's rotation, and a further effect will be to decrease the Moon's distance from the Earth and thereby to diminish the Moon's orbital period. The Earth's rotational period will thus be greater than the Moon's orbital period—the 'day', in fact, will be longer than the lunar month—and, if the human race has contrived to survive so far, our remote descendants will have the spectacle of seeing the Moon rise in the west and set in the east, with the Sun still following its present routine of rising in the east and setting in the west. The Moon will continue its approach towards the Earth until, eventually, it arrives within the danger zone (referred to on p. 85), whose outer limit is about 10,000 miles from the

Earth's centre, when the gravitational strains imposed by the Earth's attraction on the solid Moon, will cause the Moon to be broken up into fragments, later forming a ring such as we see around Saturn.

It is not to be supposed that the Earth itself will remain unscathed during the catastrophic progress of the Moon. There will be collisions amongst the lunar fragments and undoubtedly some of these will fall on the Earth as giant meteorites, possibly as large as minor planets. But long before the Moon comes within the danger zone the fate of the Earth will have been effectively decided; if the terrestrial oceans still exist the lunar tides will rise to enormous heights and at intervals of a few hours will sweep over the surface of the Earth, engulfing most of the land surface in their progress; perhaps the only areas safe from the devastating tides will be the polar regions. This is a gloomy prospect for terrestrial life in the very distant future and it seems remarkable that a forecast of the Earth's fate, unless some counterbalancing agency intervenes, has its origin in what seems such a trivial cause, the present minute amount of tidal friction in the shallow seas.

In the two previous chapters we have seen that the geological evidence points to an assignment of at least 2000 million years for the age of the Earth; the ages of the oldest meteorites are perhaps 50% greater than this. The astronomical evidence discussed in the present chapter also indicates an age for the solar family of the same order. All our lines of inquiry accordingly converge in a remarkable way in enabling us to answer the question 'When?' with considerable confidence, although not with the spurious precision of Archbishop Ussher. We may put the age of the Earth and the Solar System at something like 3000 million years—it may be somewhat greater or somewhat less—; but whatever its precise value it would appear that a very few thousands of millions of years ago the Sun's family of planets came into being as an organized system. Beyond that early date in planetary history and in the history of the Universe is mystery, unfathomed in the present state of our knowledge and perhaps unfathomable for all time; for here we are approaching the moment of Creation, not presumptuously because of the achievements of science but in a spirit of true reverence for the Power that transcends all human intelligence.

Part III

HOW?

CHAPTER IX

FROM THE NEBULAR HYPOTHESIS TO THE PRESENT DAY

WE now come to the third of our questions, 'How?' We have just seen that we can answer our second question 'When?' with considerable and even surprising success. The answer to our first question 'Whence?' was in one sense definite since we came to the important conclusion that the planetary family must have come into existence as a consequence of some single process; but, it will be recalled, we were unable to determine the exact parentage of the planets, for it was suggested that the Solar System might be the product either of the Sun's own unaided efforts or of the interaction of the Sun with another star. In this final section of the book we consider some of the processes which have been suggested from time to time in answer to the question, 'How?'

It must be said at once that we are entering the field of speculation at this stage of our inquiry; but, so far as is possible, we do not depart from the principles, inherent in the scientific method, of attempting to arrive at a given result—in this connexion, the present state of the Solar System—from an *assumed* event, or set of circumstances, that occurred or were operating in the distant past. The method, then, is one of 'trial and error'; it involves, first, the statement of a hypothesis which must of course be reasonable so far as our present knowledge of the Universe is concerned and, second, the display of the whole chain of arguments flowing from such a hypothesis. The element of speculation arises from the fact that it is not always possible to work out the detailed implications of the hypothesis with anything like the confidence that is found, for example, in the calculation of the ages of the rocks by the uranium-lead method; the individual problems which are encountered are, in fact, too complex to be resolved mathematically and it is then only possible to indicate qualitatively or even to guess what results may flow from a given event or set of circumstances.

The various processes that have been suggested for the formation of the Solar System can be broadly divided into two categories. Just as a century or more ago the geologists were arrayed in two camps, one uniformitarian and the other catastrophist, so cosmogonists—past and present—are divided into two groups, one disposed to accept the hypothesis that the Solar System is the result of a gradual evolutionary process, and the other prepared to believe that some cataclysmic action—usually associated with the hypothetical encounter of the Sun with a star in the distant past—must be invoked.

In recent years there has been a considerable addition to the number of theories purporting to account for the origin of the Solar System and this fact alone is sufficient to suggest to the reader that nothing approaching finality, in the shape of a reasonable and acceptable theory, has so far been reached. It is quite possible that we shall never know, beyond a shadow of a doubt, how the planetary system came into existence; perhaps the most that cosmogonists can hope to achieve is the formulation of a theory that is, first, in full accordance with established physical principles, second, within the bounds of probability as regards the process envisaged and, third, capable of accounting for the general features of the Solar System as we know it at present. In describing some of the outstanding attempts in this field of scientific inquiry we shall not, as a rule, indulge in detailed criticisms of all the aspects of the implications of the theories concerned, for many of these would take us to highly technical levels; our main aim will be to give a general description of some of the ideas involved in each theory and to indicate, usually briefly, where a particular theory succeeds and where it fails.

At first sight it is not immediately evident that the second type of hypothesis based on the interaction of the Sun with a star should be invoked at all, since the Sun appears to be so completely isolated from other stars in the Galactic System. The star nearest to the Sun is Alpha Centauri which is about 25 million million miles away or, expressed in light-time, about $4\frac{1}{3}$ light-years away. We may suppose that this is roughly the average distance between any star and its nearest stellar neighbour; * and if we remember further that the average star has a diameter of perhaps one million miles, it is evident that galactic space is mainly emptiness—at least so far as the stellar population is concerned.

It is difficult to visualize the isolation of the Sun from its stellar neighbours, but perhaps the following illustration will be of some assistance; if we represent the Sun by a marble in London, the stars in the immediate vicinity of the Sun will be represented by marbles of varying sizes as far away as Edinburgh, Dublin, Paris, Lyons, Brussels, Copenhagen and so on. But even this picture is incomplete, for we have to remember that the stars are not stationary but are moving with considerable speeds in galactic space. It is found that the Sun is travelling, with reference to the stars in its neighbourhood, with a speed of approximately 12 miles per second nearly in the direction of the bright star Vega; with this speed the Sun would require about 70,000 years to travel a distance equal to the average spatial separation of the stars, and on the scale of our illustration the marble representing the Sun would be moving with a speed of but an inch per day, perhaps in the direction of Moscow!

Like the Sun all the stars are in motion, with all sorts of speeds in all sorts of directions, and when we envisage the Galactic System as a whole we interpret the stellar motions as motions in immense orbits described around the centre of the Galaxy. This, as we have seen (p. 181), is the basic idea underlying our conception of the phenomenon known as the rotation of the Galaxy. When we remember the vast distances

* We except, of course, components (that is, the individual members) of double or multiple stars.

separating neighbouring stars and the comparatively minute sizes of the stars themselves it is evident that, despite the high stellar speeds involved, the chance of a star making a close approach to another star—say, the Sun—is extremely small. If we suppose that a star *A* just grazes a second star *B* at some time or other, then the average interval between this encounter and a subsequent grazing encounter of *A* with another star *C* is about 50,000 million million years.

When we recall that the age of the Sun and, presumably, the other stars in the Galactic System is of the order of 3000 million years, it is immediately evident that the probability of the close approach of one star to another is indeed very small; accordingly, if such an encounter is postulated in connexion with the formation of the planetary system, we must infer that such an event is certainly very rare and may even be unique so far as the Galactic System is concerned. It should be added that the previous arguments are based on the present sizes and distribution of the Galactic stars accessible to our telescopes; if the stars were very much larger, as they might quite well have been several thousand million years ago, and if the Sun in the course of its orbital motion around the galactic centre passed through the dense concentration of stars in the neighbourhood of the galactic centre, the probability of the Sun's encounter with another star would be considerably increased, perhaps to the extent that an encounter of this sort would not be quite the uncommon event among the galactic stars, as is generally supposed. Whatever be the degree of probability or improbability of close encounters we have the undoubted fact of the existence of the Solar System and, if all other hypotheses as to its origin fail, it is legitimate to inquire if our planetary system could be brought into being through the close approach of another star to our luminary.

Until 1942 there was no evidence of the existence of bodies of planetary mass outside the Solar System. In that year it was announced that several stars were accompanied by bodies that might conceivably be large planets as *distinct from stars*. The arguments leading to this provisional conclusion may be illustrated in terms of the motion of the Earth-Moon system around the Sun. We ordinarily say that the Earth's orbital path around the Sun is an ellipse; more accurately, we ought to say that the centre of mass of the Earth and Moon moves in an ellipse around the Sun. Now, since the Moon describes

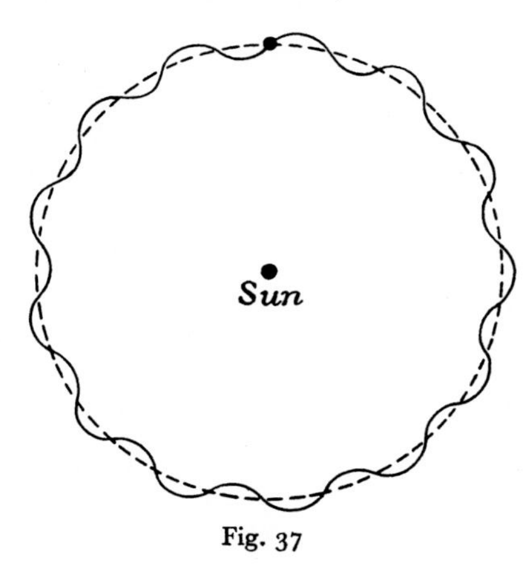

Fig. 37

a path relative to the Earth in a period of $27\frac{1}{3}$ days, both the Earth and the Moon will describe paths relative to the centre of mass in the period just mentioned. Suppose now that we have an astronomer on a nearby star and that his instruments are sufficiently powerful to enable him to see the Earth but not the much feebler Moon.* He would find from his observations that the path of the Earth around the Sun during a terrestrial year would be a wavy line as represented in Fig. 37. Basing his interpretation on the law of gravitation he would infer that the Earth was one member of a twin

* If he was equipped with an instrument as powerful as the new 200-inch telescope on Mt Palomar, he would actually be unable to see Jupiter (the brightest of the planets) and, of course, the Earth would also be invisible.

system, the second member being invisible; further he would be able to deduce the path of the centre of mass of the system around the Sun (the broken line in Fig. 37) and the ratio of the masses of the Earth and Moon.

This is essentially the process adopted in the stellar case. One of the stars concerned is a double star known as 61 Cygni,

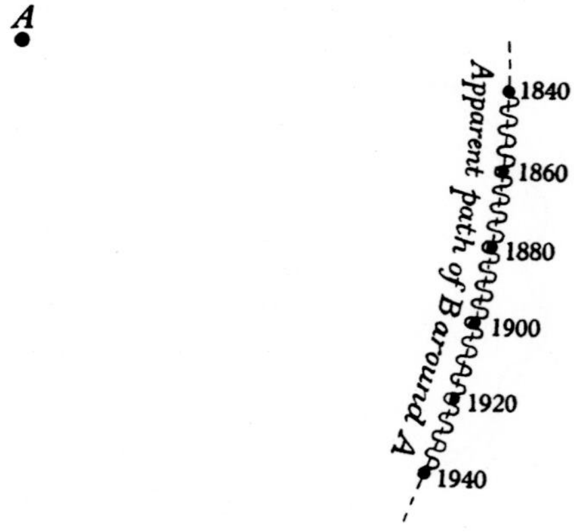

A

Apparent path of B around A

1840

1860

1880

1900

1920

1940

Fig. 38

famous in astronomy as the first star to have its distance measured successfully; it consists of two members, or components—we refer to them as *A* and *B*—bound together by the law of gravitation in much the same way as the Earth-Moon system is held together; observations over a series of years enable us to plot the apparent path of *B* around *A*. Until recently this path was supposed to be part of an ellipse, shown by the broken line in Fig. 38. But more accurate observations have now shown that the path of the luminous star *B* around *A* resembles in character the wavy line in Fig. 38; in this figure the broken line represents part of the elliptic orbit of *B* around *A* when the observations are smoothed over intervals of about ten years, thereby concealing the fluctuations with periods of about three years;

the wavy line is extended back to 1840 although the reality of the fluctuations was not suspected till a century later; the fluctuations are not drawn accurately and in the figure are intended to give some idea of their character only. It is finally inferred* that either *A* or *B* is accompanied by an unseen body *C* whose mass is deduced to be about one-sixtieth of the mass of the Sun (or about seventeen times the mass of Jupiter). The smallest stellar mass measured is not much more than one-tenth of the mass of the Sun and the star concerned is a very feeble luminary. If the mass attributed to the unseen body *C* is reasonably accurate, it may be permissible to describe *C* as a planet rather than as a star.

The unseen bodies accompanying the other stars concerned prove to be somewhat more massive than the body *C* that we have just been discussing and are perhaps close to the rather vague borderland separating stars of very feeble radiating power and the non-luminous bodies with which the name of 'planet' is associated. Unless it is proved eventually that such an unseen body as *C* in the system of 61 Cygni is but one, although the most massive, of a retinue of planets belonging to the star concerned, it would seem that the explanation of the origin of such bodies is to be sought in the process accounting for the formation of double stars rather than in the process responsible for the formation of a planetary system. In any event we must suspend judgement until much more information is available.

The Nebular Theory of Laplace

We shall first consider in some detail one representative theory of the uniformitarian class and one theory of the cataclysmic

* The path of *B* relative to *A* is not fundamental; we could just as easily use the observations to derive the path of *A* relative to *B* and the wavy characteristics would be shown up in a similar way. It may be added that the reliable observations extending over the last one and a half centuries are sufficient to cover only a part of the complete orbit, for the orbital period is approximately 700 years.

type, for all other theories of the origin of the Solar System have something in common with one or other of the types mentioned; we begin with the nebular theory of Laplace.

This celebrated theory was published in 1796 and for a century it enjoyed a unique popularity in cosmogonic speculations. The theory was evidently prompted, or at least influenced, by two considerations: first, the existence of nebulae in the Universe, many of these having been discovered through the observational activities of Sir William Herschel and, second, the well-known ring-system of Saturn (Plate II*c*, facing p. 17). The nebulae belong mainly to two classes, (i) the extra-galactic or spiral nebulae of which the Great Nebula in Andromeda (Plate VII, facing p. 167) is perhaps the best-known example, and (ii) the diffuse gaseous nebulae, of which the Great Nebula in Orion is an outstanding example; the known nebulae of the second class are at moderate distances from the Solar System and are situated in or close to the Milky Way. Comparatively recent research has shown us that the nebulae of the first type are, in most respects, replicas of the Galactic System, that is, vast aggregations of stars, diffuse nebulae and star-clusters, at immense distances from us; for example, the Andromeda nebula, one of the nearest of these objects, is at a distance of 700,000 light-years while the most remote so far surveyed, as already mentioned, is 500 million light-years distant (about three thousand million million million miles). Such information was of course not available in the time of Laplace (one and a half centuries ago) and, accordingly, we must suppose that his hypothesis related to nebulae of the second class.

As the basis of his theory Laplace supposed that far back in time the Sun had been a great gaseous globe, or 'nebula' as then conceived, consisting of a comparatively dense nucleus beyond which extended an immense atmosphere reaching to a distance exceeding that of the farthest planet then known, namely Uranus; in our account we shall suppose that the 'nebula' extended well beyond the orbit of Neptune (we ignore Pluto in this connexion). Also, Laplace introduced the further hypothesis that the 'nebula' was originally rotating. The sequence of events was then supposed to be as follows. Under the general gravitational attraction, the globe would slowly contract and, in accordance with a well-established law of dynamics, its rotation would become more rapid. Just as mud is thrown off the rim of a rotating wheel when the rotation is sufficiently rapid—in this case the centrifugal force exceeds the force of adhesion tending to hold the mud to the rim—so Laplace supposed that when the centrifugal force in the outer layers of the 'nebula' exceeded the gravitational attraction of the 'nebula' as a whole, gaseous matter was thrown off, later to form a ring, like Saturn's Ring, revolving in the equatorial plane of the 'nebula' in the same direction as that of the nebular rotation. The 'nebula' continued to contract while the material of the ring was supposed to be slowly collected into a single aggregation of gaseous matter which, on further condensation and cooling, developed into a planet revolving around the central body in approximately a circular orbit lying in the plane of the solar equator. As the result of the further shrinkage of the 'nebula' and of its increased rotation, a second ring was in due course thrown off, later to become a planet by the process already described.

In this way Laplace attempted to account for the formation of the seven planets known in his day and for the common direction of revolution around the central body in much the same plane and in orbits which were roughly circular, all in accordance with the observed uniformities of the planetary system. It should be remarked, however, that the plane referred to should be, substantially, the plane of the Sun's equator; but the plane which best expresses the actual orbital planes of the planets, the 'invariable plane' as it is called, is inclined at 6° to the plane of the solar equator. Laplace attempted—unconvincingly, as it now appears—to explain a further regularity, namely, the common direction of rotation of the planets. The satellites were supposed to be formed from the parent planet, at a time when it was still an extended gaseous globe subsequent to its formation from the nebular ring, by a process similar to that we have described concerning the formation of the planets from the original 'nebula'.

It may be interpolated here that the real originator of the nebular theory was Immanuel Kant, the famous German philosopher, whose book *A General Theory of the Heavens,* or *Essay on the Mechanical Structure of the Universe, on the Principles of Newton* was published anonymously in 1755 when the author was a *Privatdozent* at Königsberg University engaged in lecturing on mathematics and physics. The chief differences between Kant and Laplace were: (i) Laplace assumed that the original state of the 'nebula' was one of rotation, whereas Kant attempted to show how rotation could be started, and (ii) Kant was not influenced, as Laplace was, by the supposed analogy with Saturn's Rings.

Despite its apparent success in accounting for several of the chief features of the planetary system, the nebular theory is to-day completely discredited, so far at least as the formation of the planets and satellites is concerned. It may be added that there is some justification for applying Laplace's hypothesis to the formation of *stars* from a vast rotating nebula for, as Jeans has emphasized, the various types of extra-galactic nebulae seem to indicate stages in the evolutionary process of the formation of stellar systems from diffuse matter. Laplace's appeal to the analogy of Saturn's Rings ante-dated by more than half a century Clerk-Maxwell's famous research on the stability of the rings; this investigation made tolerably clear that the rings, however constituted, could never be collected into a planet as Laplace supposed. On the contrary, the most satisfactory explanation of Saturn's Rings is precisely the opposite, namely, that they are the debris of a large satellite which somehow got within the danger-zone where the planet's gravitational attraction was sufficient to disrupt the satellite into an immense number of fragments, the real constituents of the rings.

With our present knowledge Laplace's suggestion as to the formation of satellites can hardly bear scrutiny. At birth, a satellite must be gaseous, at a considerable temperature and greatly distended beyond its dimensions in its solid condition. Under these conditions most satellites would be unable to retain any gases and vapours and they would simply 'melt away', their substance being dissipated throughout interplanetary space. It has been calculated by Jeffreys that if a satellite (or, for that matter, a planet), composed mainly of rock, has a diameter at present less than about 2500 miles, then it could not be formed through the condensation of a gaseous mass. Most of the satellites have diameters much smaller than that just mentioned, and it seems certain that the great majority of the satellites could not have come into existence by the process described by Laplace.

There is a further serious objection to the nebular theory; this relates to the consequences of the initial assumption that the Sun was originally distended beyond the present orbit of the outermost planet. Immense giant stars are of course known; for example, the diameters of two of the largest stars known, VV Cephei and Epsilon Aurigae are about 1100 million miles and 1600 million miles respectively, which, however, are small in comparison with the Sun's hypothetical diameter of 6000 million miles if it was distended to the distance of Neptune's orbit. We have seen that the age of the Sun can be reckoned as a few thousand million years and in this time its mass could not be diminished much by radiation; further, the geological and biological records are emphatic that the Sun could not have changed much in other respects during the period of time associated with the oldest terrestrial rocks. It would then seem unlikely that the Sun has altered substantially during its lifetime. If then the planets are condensed out of rings thrown off, as Laplace suggested, by a rotating globe whose dimensions were little different from the present dimensions, how could the planets have been pushed out to their present immense distances? This has seemed until recently to be an insuperable objection to the nebular theory; however, we shall see later that there is a possible loophole of escape from this particular dilemma, although on other grounds the hypothesis of a gigantic sun at the beginning of things breaks down completely, as we now try to show.

Let us suppose, then, that originally the Sun was distended and endowed with rotation as Laplace supposed. To follow out the consequences of this hypothesis we invoke a dynamical principle known as the conservation of angular momentum. As applied to the nebular hypothesis this principle states that the angular momentum of the original Sun is equal to the angular momentum of the present Sun together with the angular momenta of the planets.* The concept of momentum involves mass and velocity and as applied to a planet such as Jupiter, which we assume for simplicity to be describing a circular orbit, the orbital momentum of Jupiter with reference to the Sun's centre is the product of its mass and velocity; the angular momentum of Jupiter is the product of the momentum, as defined, and the radius of the planet's orbit. Theoretically, we ought to take into account the contribution to the total angular momentum resulting from the planet's axial rotation, but as this is small we can ignore it in the present connexion. In the case of the Sun's axial rotation the angular momentum is the sum of the angular momenta of all the particles composing the Sun, and this sum depends principally on the Sun's mass, the way this is distributed within the Sun, and the magnitude of the solar rotation. It is found that of the total angular momentum of the Solar System as at present, the contribution of the planets amounts to 98% and the Sun's to only 2%, a distribution that is entirely inconsistent with the principle of conservation of angular momentum

* It is to be remarked that the Sun and the nebula depicted in Fig. 39, being gaseous, do not rotate as rigid bodies.

if Laplace's theory is invoked. We examine the argument in simple terms as follows.

In Fig. 39 (*a*) we picture the original rotating nebula N_1; the outer shell, shown in light shading, is the part that is about to break away to form the first ring. Fig. 39 (*b*) shows the ring and the remaining nebula N_2 just after the ring has been formed. Fig. 39 (*c*) represents the state of affairs when the outermost planet has been formed, as Laplace imagined; the planet's orbit is indicated and the nebula is now repre-

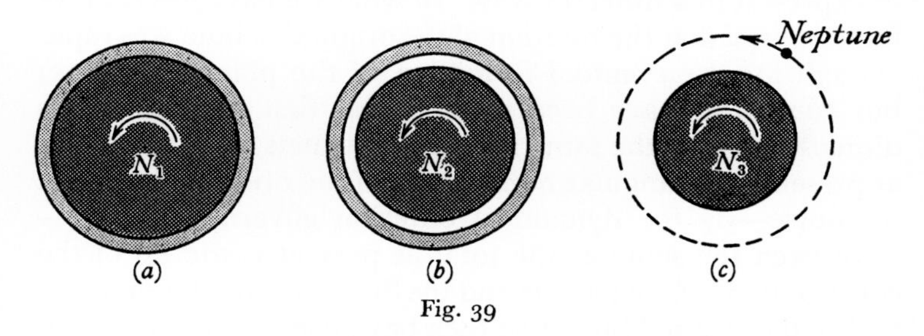

Fig. 39

sented by N_3, somewhat more condensed and rotating more rapidly than N_2. By the angular momentum principle, the total angular momentum of N_1 is equal to the sum of the angular momentum of the ring, in (*b*), and the angular momentum of the nebula N_2, for the angular momentum of the ring is substantially the same as the angular momentum of the shell shown shaded in (*a*). But, as calculation shows, the angular momentum of the shell can be but a very small fraction of the angular momentum of the original nebula N_1, for the mass of the shell, if identified with the mass of Neptune, is but a twenty-thousandth part of the mass of N_1. It follows that in (*b*) the angular momentum of the ring is but a very small fraction of the angular momentum of the nebula N_2. Further, the angular momentum of the ring in (*b*) will be concentrated into the angular momentum of Neptune in (*c*); consequently, the angular momentum of Neptune will be but a very small fraction of the angular momentum of the nebula N_3.

The same arguments apply to the formation of the other planets. It is thus seen that the nebular theory requires that only a small part of the angular momentum of the original nebula should be found in the planets and that the most of it should be found in the Sun. As we have seen, the distribution of angular momentum is, in fact, precisely the reverse and we must conclude that the process envisaged by Laplace is impossible.

The argument is so important that it may be worth while to express it in a different way. In what we have just said we have *assumed* that the rotation of the original nebula was rapid enough for the eventual formation of the planet Neptune; but could this have been so? Suppose first that the Sun's dimensions were the same before the planets were formed as at present. The angular momentum of the original Sun must of course—by the dynamical criterion governing events—have been the same as the total at present made up of the contributions of the planets and the Sun; the angular momentum of the original Sun must then have been about fifty times greater than that of the Sun at present, from which it follows that the original Sun must have been rotating about fifty times faster than at present; its period of rotation must then have been about 12 hours (the present period is about 25 days), that is, just a little greater than the rotational periods of Jupiter and Saturn. The effect of this more rapid rotation would be to give the Sun a spheroidal shape—not unlike the shape of Jupiter and Saturn. But even with this rapid rotation the centrifugal force at the Sun's surface would only be a small fraction of the corresponding gravitational attraction and the throwing-off of a ring would be impossible, just as the throwing-off of mud from a rotating wheel would be impossible if the centrifugal force (depending on the rotation of the wheel) were very much smaller than the force of adhesion between the mud and the rim of the wheel. If we now suppose that the original Sun is distended to the orbit of Neptune, the rotation would be very much slower and although the gravitational force at the surface would also be much smaller, the result would be again the impossibility of matter being thrown off to form a discrete ring. We have gone into the arguments in some detail, which the criterion of angular momentum demands, for in other theories it has to be invoked.

Tidal Theory

The formidable objections to the nebular theory dispose of the simplest form of the uniformitarian hypothesis which possibly can be rescued from failure only by the introduction of new ideas and new processes. It was accordingly natural for cosmogonists, on the exposure of the weaknesses of the Laplacian theory, to turn to the alternative idea of a cataclysmic origin of the planetary system.

We describe at this point the principal features of the tidal theory put forward some years ago by Sir James Jeans and modified by Professor Harold Jeffreys. In this theory the formation of the planetary system is attributed to the results of the interaction with the Sun of a star passing close to our luminary. Just as the Moon by its gravitational attraction raises tides on the terrestrial oceans, so the star, assumed to be much more massive than the Sun, raises immense gaseous tides on the Sun; in effect the Sun's shape is distorted, as illustrated in Fig. 40. Moreover, matter is pulled away at *A* from the Sun, roughly in the direction of the star, to form a long gaseous filament—as represented in the figure—which, as the star recedes, will be extended to greater distances from the Sun and, at the same time, given a rotary motion around the Sun. The filament, according to Jeans, is unstable lengthwise and it will soon break up into several sections each forming a distinct aggregation of matter later to develop by cooling and contraction into a planet. Jeffreys suggested, however, that the aggregations may come into being shortly after the ejection at *A* of the corresponding solar material; in this sense the filament is no longer regarded as continuous (as shown in Fig. 40) but as a series of condensations whose masses and spatial distribution are calculated to be roughly in accordance with the known characteristics of the planets in the Solar System at present.

The result of the stellar encounter is then a number of aggregations of planetary mass which are given a lateral motion around the Sun through the gravitational attraction

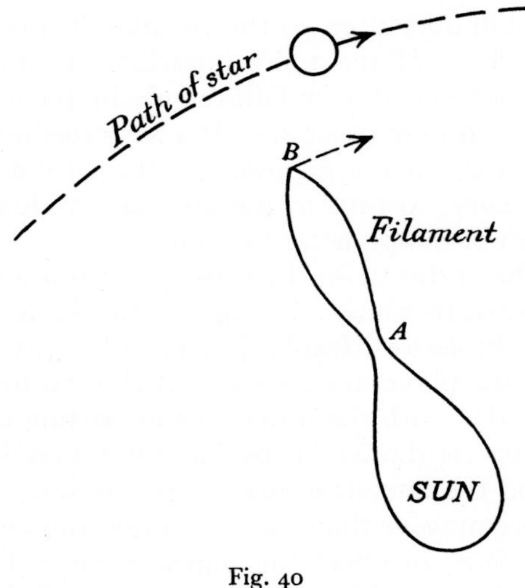

Fig. 40

of the retreating star; these aggregations, which soon condense into planets, revolve at first around the Sun in highly elongated orbits, all described in the same direction and all much in the same plane; this plane would be the plane of the star's relative motion with respect to the Sun and, if the Sun were originally rotating, the plane of the planetary orbits could hardly be expected—except as a most improbable coincidence—to be the same as the plane of the solar equator. Accordingly, the difference of 6° (previously mentioned) between these two planes is *not* an argument against the tidal theory, although it does constitute a definite argument against the Laplacian theory.

Consider now one of the planets formed in the way indicated. After one or more revolutions it will again be close to the Sun and the tidal action of the Sun on the planet will result in the partial disruption of the latter with the formation of satellites, the process being similar to that by which the planets come into being as a result of the tidal action of the visiting star on the Sun. Since the satellites, as a class, are too small to retain an atmosphere especially at considerable temperatures, they cannot be gaseous at birth; it must then be assumed that when this event occurs the parent planet must have cooled effectively to the liquid state. This explanation of the formation of satellites does not apply to the exceptional case of the Moon; the resonance theory of the Moon's birth is described on p. 185.

As stated, the original orbits of the planets according to the tidal theory are extremely elongated while the present orbits are nearly circular. This change in character is supposed to be brought about as follows. The star's encounter with the Sun is hardly likely to have produced such 'tidy' results as are illustrated in Fig. 40; in addition to the filament there must have been solar matter dispersed all around the Sun, forming a resisting medium through which the planets would have to plough their way. It is known that one effect of a resisting medium is to reduce the eccentricity of a planet's orbit and it is supposed that the present nearly circular character of the orbits has been produced in this way.

The tidal theory meets serious difficulties in attempting to account for the rotation of the planets. Also, the angular momentum criterion raises apparently insurmountable objections, for it is difficult to understand how the planets could ever have acquired their present amount of angular momentum; or, expressed somewhat differently, how they could have got pushed out to their immense distances from the Sun. The only loophole of escape from this dilemma is to assume that the Sun was enormously more distended even than Jeans supposed. As a result of various arguments, some of which have already been mentioned, it has been regarded until recently as axiomatic that during practically all its life the Sun's dimensions have changed very little, leaving only a relatively insignificant interval for condensation from a nebular state. But, to-day, part of the difficulty concerning the solar dimensions in Laplace's and Jeans's theories can be met by reference to the nuclear reactions described in Chapter VIII. There we indicated that the contraction of a star from a nebular condition proceeded in jumps; first, we considered the partial condensation sufficient to provide the conditions for starting the nuclear reactions involving protons and the lighter elements (it will be recollected that we paid most attention to the proton-lithium reaction); the Sun must have been greatly swollen when these reactions were occurring and it would remain distended for a very long time until the supply of the appropriate 'fuel' became exhausted; after this stage, as we have seen, the Sun would rapidly* contract approximately to its present dimensions when the conditions necessary for the carbon-nitrogen cycle were attained. Taking into account the very much greater 'target-area' presented by the distended Sun while the earlier proton reactions are taking place, it would seem that the probability of an encounter at this time is considerably greater than in the interval during which the Sun has retained its present dimensions, despite the fact that the present stage represents perhaps 80 or 90% of the Sun's age. This argument adds some weight to the suggestion that the Sun's family of planets is perhaps not the unique system as is so generally supposed.

We now consider—rather briefly—several alternative theories most of which are, to some extent at least, basically related to one of the two theories already described. Much of the interest in these theories resides in the novelty of the ideas introduced and this is the aspect with which we shall mainly be concerned.

* In the 'Kelvin-interval' of one or two scores of millions of years.

The Chamberlin-Moulton Theory

Allied in some measure to the tidal theory is the earlier 'planetesimal theory' advanced by Chamberlin and Moulton in 1905. This theory was suggested by the solar eruptions known as 'prominences'; the Sun is by no means a quiescent body, and frequently gases and vapours are expelled to great heights above the general level of the solar surface; these eruptions, which can now be studied exhaustively whenever they occur by means of an instrument called the spectroheliograph, are called 'prominences'. It was supposed that in the distant past the intensity of these eruptions was on a vaster scale than at present and that when a wandering star came close to the Sun the gases and vapours of the prominences were dragged out, through the gravitational attraction of the star, to immense distances from the Sun when, on cooling, they condensed into small solid bodies called 'planetesimals'; it was then supposed that the planetesimals were collected into several main aggregations which later became the planets. Although the theory breaks down, in particular, on the assumption that a comparatively small amount of gas or vapour can condense into a so-called planetesimal—it is certain that such an amount of gas or vapour would simply be dissipated into space—the new ideas helped to reawaken interest in the planetary problem and to stimulate further investigation in this engrossing subject.

The Collision Theory

The failure of the tidal theory to afford a satisfactory explanation for the rotations of the planets suggested to Jeffreys the revival of an earlier hypothesis that the star's encounter with the Sun took the form of an actual collision supposed, however, not to be exactly 'head-on'. Owing to the difficulties of submitting this hypothesis to mathematical treatment, the supporting arguments are mainly of a qualitative kind. Nevertheless, the theory has its successes, particularly as regards planetary orbits and rotations. Since the dimensions of the Sun at the time of the collision are supposed to be much the same as at present, the angular momentum criterion seems to raise insuperable difficulties, as in the case of the tidal theory, in accounting for the immense distances of the planets from the Sun. It is not clear that the collision hypothesis would gain as a whole if the Sun were imagined to be very greatly distended at the time of the collision, although the probability of such an event would be considerably enhanced and the angular momentum difficulty would be minimized.

The Binary Star Hypothesis

As we have seen, the chief argument against the tidal and collision theories—and the nebular theory as well—is the impossibility of suggesting a reasonable process whereby the planetary matter is removed from the immediate neighbourhood of the Sun to the present great distances of the planets and set in motion in nearly circular orbits. The difficulty would be immensely reduced if the matter of which the planets are formed was originally as far away from the Sun as Saturn, for example, is at present; in this event the planets cannot be the offspring of the Sun itself. It occurred to Professor H. N. Russell that if the Sun was at one time not a single star but a binary star (that is, a twin-star system) and that if another star made a close approach to the Sun's companion (the other member of the twin system) producing such tidal effects as we have already described, then the angular momentum criterion would be satisfied to a great extent and the disabilities of the original tidal theory would be substantially diminished. The suggestion that the Sun was originally a binary star, or more accurately one component of a twin system, causes little difficulty, for binary stars are anything but uncommon in the Galactic System. According to the most conservative estimates, at least 10% of the stellar population * consist of binaries, divided into the three categories—

* Some authorities put the percentage as high as 30%.

depending on the methods of detection—of visual binaries, spectroscopic binaries and eclipsing binaries. A visual binary, as its name implies, is actually seen in the telescope to consist of two stars, referred to as 'components', one of which appears to circulate around the other, just as the Moon circulates around the Earth, according to the law of gravitation; the orbital periods of such stars vary from about a year to hundreds or even thousands of years. For example, the period of the binary star Sirius is about 50 years and of 61 Cygni about 700 years, as we have already mentioned. The smallest distance between the components of a visual binary is about 75 million miles; for 61 Cygni the distance is roughly 8000 million miles, about three times the distance between the Sun and Neptune; for stars of longer period the distance is still greater. As regards the spectroscopic and eclipsing binaries the periods are generally a few days only and the distances between the components vary from one or two million miles upwards. Binary stars, as a class, thus provide a wide range of variety.

Russell's suggestion of a stellar encounter with a binary Sun raises the very real difficulty of accounting for the removal of the Sun's companion from our luminary's control and for the retention of the tidal filament which is later supposed to condense into planets as in the tidal theory. An ingenious solution of this problem was given by Dr R. A. Lyttleton in 1936. The circumstances of the encounter are illustrated in Fig. 41. It is supposed that the orbit of the companion relative to the Sun is comparable in size with the orbit of Saturn or Uranus; as an example Lyttleton takes the radius of the orbit to be about 1700 million miles so that, if the companion's mass is assumed equal to the Sun's mass, the period of the binary is about 50 years and the companion's speed around the Sun is about 6 miles per second. The path of the intruding star relative to the Sun, not necessarily in the plane of the binary, is shown in the figure; its speed relative to the Sun is assumed to be at least 20 miles per second. So far the hypotheses are entirely reasonable and unexceptionable.

To produce tidal effects of sufficient magnitude, the intruder must pass close to the companion, the minimum distance being supposed to be three or four million miles; as the intruder is always at a comparatively great distance from the Sun, its tidal effect on the Sun is negligible. The tidal

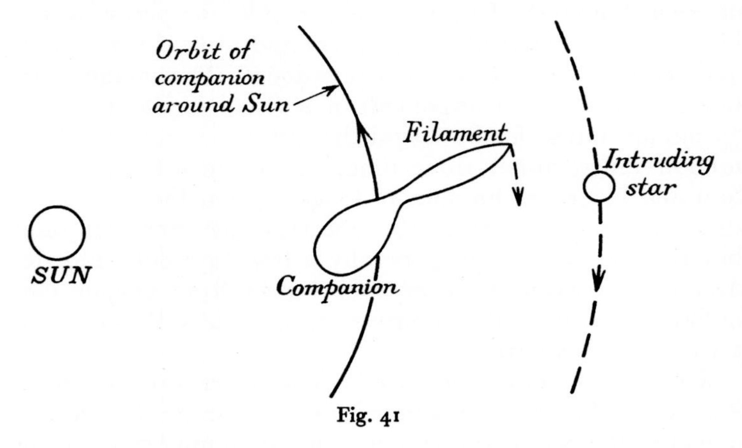

Fig. 41

filament ejected from the companion is shown diagrammatically in Fig. 41. If the mass of the intruder is comparable with the mass of the companion, it is more than probable that a tidal filament, due to the companion's gravitational attraction, will be ejected from the intruder as well. At first sight it is difficult to understand how the companion's elliptic orbit around the Sun can be converted into a hyperbolic orbit—a necessary condition if the companion is to escape from the Sun's gravitational control—and how *at the same time* the filament can pass substantially under the Sun's control, later to condense into planets. Lyttleton, however, showed that within comparatively wide limits pertaining to the initial conditions this course of events was possible; the companion—and the intruding star—would disappear into galactic space and some parts, at least, of the filament would move in elliptic orbits around the Sun. It is to be noted that all the matter forming the planets is originally within a small volume of space and, when the planets are formed by condensation as in Jeans's theory, they will continue, for some time at least, to pass very near to one another; in their earlier encounters the planets must be expected still to be fluid and tidal action between any two planets in this condition at a time when they are close together is supposed to result in the formation of the satellite systems.

The theory accounts satisfactorily for the large preponderance of angular momentum associated with the planets—as it must do, of course, since the initial hypothesis was designed to ensure this specific result. It is open, however, to several objections one of which must immediately occur to the reader; if the planets are all formed at much the same distance from the Sun, what process has supervened to impose the immense changes in their distances from the Sun to bring them into their present orbits? It must be concluded that, despite its ingenuity, the binary hypothesis in the form stated above and in its subsequent modifications is no nearer a tolerably satisfactory solution of the origin of the planetary system than its predecessors.

The Fission Theory

Although we are not primarily concerned with the evolution of binary stars, the process by which it is believed they come into being is utilized in a theory of the origin of the solar family recently propounded by Ross Gunn. If a star is rotating originally, the rate of rotation will increase as and when contraction proceeds, in accordance with the principle of the conservation of angular momentum. If the rotation becomes exceedingly rapid, the stability of the star is imperilled and eventually the star breaks up into two parts, the components of a binary star. This process is known as 'fission' and it is supposed that the close binaries have been formed in this way. Now the principal objection to the earlier theories, such as the tidal theory, was concerned with the problem of getting the planetary material to sufficiently great distances from the Sun, for the intruding star had, so to speak, to do all the work of tidal ejection; if, however, the Sun itself could make a substantial contribution to this last process, the objection would not be so serious.

Gunn's suggestion is that an encounter takes place between a rapidly rotating star and another; the former is on the point of breaking up and so has very little gravitational control over part of its material, with the result that the tidal effect of the intruding star is very greatly enhanced. Without going into further details it is sufficient to say that this theory strains the probabilities excessively; as we have seen, an ordinary encounter must be regarded as a most uncommon event, but when one of the stars involved is further supposed to be just on the point of parting with its last modicum of stability so that the process of fission is almost due to begin, the supposed event can hardly be conceived as a real possibility.

The Cepheid Theory

The idea of assuming that one of the stars in an encounter can be rendered unstable is the basis of a theory advanced by Professor A. C. Banerji in 1942. He starts with a Cepheid variable, such as Delta Cephei with which we were concerned in pp. 159–166. A star of this type pulsates rhythmically and, if left to itself, it would continue to pulsate—at any rate, over a long interval of time—without any danger to itself. But an intruding star will change the situation very definitely; the extent of the pulsations will increase, almost without limit, under the gravitational attraction of the intruding star and the Cepheid will become unstable in much the same general sense as in Gunn's theory; it is accordingly surmised that it will throw off vast quantities of matter* some of which, it is

* It is to be remembered that Cepheids are from five to twenty times more massive than the Sun.

conjectured, will condense into the Sun and some into planets. If the velocity of ejection is sufficiently great, the Sun and planets will part company with the parent Cepheid. The orbital motions of the planets around the Sun are supposed to be due to the lateral attraction of the intruding star which in due course disappears from the scene of action. Although all the implications of the basic suggestion have not yet been fully worked out, it is not easy to see how all the observed uniformities of the Solar System can be explained in satisfactory detail.

The Nebular-Cloud Theory

In this theory, due to Dr von Weizsäcker and published in 1944, we get back to the uniformitarian type and to a form which has some affinities with the Laplacian process. As in the catastrophic theories, we do not assume that the Sun is the sole performer; the extraneous agent in the present theory is a comparatively dense interstellar cloud of gas and dust particles into which the Sun is supposed to plunge. Such an event is by no means unlikely, for galactic space is plentifully supplied with the clouds known as 'diffuse nebulae' of which the Great Nebula in Orion is perhaps the most conspicuous example of the luminous type. There are the 'dark nebulae' as well, whose presence in the Galaxy is inferred from the blanketing effect on the distant stars lying beyond them; the famous 'Coal-Sack Nebula' is a well-known example of this type. These nebulae are generally very extensive; if the Sun passed into one of them, it would remain within the nebula for hundreds of thousands of years at least, and owing to its predominant gravitational attraction would gather great quantities of nebular material into a vast solar envelope. According to Weizsäcker the envelope would develop slowly, as a result of frictional forces, into a disk-like form with a diameter comparable with the present diameter of the Solar System and with a thickness of perhaps two or three hundred million miles. It is from condensations in this nebular disk that the planets are supposed to be formed eventually. Weizsäcker also discusses the formation of satellites. This account gives but the bare bones of the theory; a more detailed account would involve us in highly technical matters. It may be said, however, that the theory has some promising features, notably that which accounts satisfactorily for the present distances of the planets from the Sun. And it may be added further that if the solar family is produced in this way, our planetary system is likely to be but one of many in the Galaxy, for the passages of stars through nebulae can hardly be regarded as uncommon events even at present;* and in the distant past, when condensation of stars out of nebular matter had only proceeded some way, the passage through one or more nebulae would be almost the normal experience of any star.

The Electromagnetic Theory

In all the previous hypotheses described, only mechanical forces such as gravitational attraction and friction have been called on to explain the formation of the planetary system. A new departure has been made recently, in 1942, by Dr Hannes Alfvén in summoning the assistance of electromagnetic forces. It is to be first remarked that, like the Earth, the Sun has magnetic properties associated with it—we refer these to the Sun's 'magnetic field'—and it is this fact that forms the starting-point of the investigation. To illustrate the relative importance of the electromagnetic and the gravitational forces on an electrically charged particle Alfvén considers the magnitudes of these forces acting on a proton (the positively charged nucleus of the hydrogen atom) moving around the Sun in the Earth's orbit and he finds that the electromagnetic force exerted on the proton is 60,000 times that of the solar gravitational attraction; if a proton is moving in Pluto's orbit, the ratio is calculated to be about 250 to 1.

* 'Hubble's variable nebula', so called, shows unmistakably a bright star ploughing its way through a comparatively dense cloud.

It is accordingly concluded that if the Sun is surrounded by electrons and an ionized cloud of the elements, the atoms of which have lost one or more of their satellite electrons and are consequently positively charged, the influence of the electromagnetic forces emanating from the Sun is predominant as compared with the influence of the force of gravitation.

Alfven has first to explain how the Sun becomes immersed in a cloud of ionized atoms. As in Weizsäcker's theory he assumes that the Sun, rotating more rapidly than at present, passes into a nebula in which the atoms are initially unionized—that is, electrically neutral. These atoms are attracted gravitationally towards the Sun, acquiring energy of motion in the process; at some critical distance, depending on the element concerned, this energy is sufficient, in collisions, to tear away one or more of the outer electrons from other atoms, and by this process he conceives the Sun to be surrounded by an immense envelope of ionized atoms extending up to planetary distances. Alfvén then goes on to apply the laws of motion for charged particles in a magnetic field, showing that matter will accumulate in the Sun's equatorial plane, mainly at distances comparable with the distances of Jupiter and Saturn from the Sun, and be set in revolution around the Sun at the expense of the solar rotation. It is then supposed that most of the atoms, gaseous or otherwise, will condense into the giant planets by a process first suggested by Professor B. Lindblad. Other atoms are attracted by the planets, some, say, by Jupiter and, if it be assumed that Jupiter has its own magnetic field, then the formation of satellites ensues by the process already described in the formation of the planets. A modification of the theory is necessary for explaining the formation of the inner planets (Mercury, Venus, Earth and Mars). Such, in brief, are the main ideas associated with Alfvén's theory which, it should be noted, has not yet reached such an advanced stage of development as to be able to parry the various criticisms which it will be called on to meet.

The Nova Theory

A *nova*, or 'new star', is a star which suddenly blazes up from insignificance to become for a brief period perhaps one of the brightest stars in the sky; after a day or two of splendour it then begins to fade away, generally to return to the insignificance from which it started. Among the well-known novae of historic times are Tycho Brahe's nova of 1572, which for some days was so bright that it was easily visible in broad daylight, and Nova Aquilae of 1918 which was temporarily one of the brightest stars in the sky. Perhaps several scores of novae flare up in the Galactic System each year, but only those comparatively near are detected. Novae have also been detected in the nearer extra-galactic (or spiral) nebulae. The cause of a nova outburst is not definitely known, but the explosive character of the phenomenon is possibly due to the sudden incidence of some nuclear reaction of such violence that the star throws off vast quantities of gaseous material and expands so greatly that its luminosity increases perhaps a millionfold; as the expelled gases cool, the luminosity diminishes and eventually the star fades away to insignificance. If the explosion is excessively violent, the nova is generally referred to as a *super-nova*.

Several novae, now long past their brief splendour, still show expanding envelopes of gas, and some of the luminous nebulae are undoubtedly the results of stellar explosions. One interesting nebula of this sort is the Crab Nebula (Plate VIII, facing p. 218) which is still expanding at the rate of about 800 miles per second, as determined from spectroscopic observations; as the angular rate of expansion can be easily measured from photographs, it is then deduced that this object is about 4000 light-years away. Associated with the nebula is a luminous star, the super-nova—or what is left of it—of the original explosion. It is estimated that this star is intrinsically about 30,000 times more luminous than the Sun, that its surface temperature is about half a million degrees

Mt. Wilson Observatory

Crab Nebula

(Centigrade) and that its radius is about twice that of the Earth; further, the total mass of the gaseous envelope is estimated to be about 15 times the Sun's mass. The known rate of expansion of the nebular shell enables us to estimate the approximate date of the appearance of the super-nova; this is found to have occurred about 900 years ago, sufficiently close to the date A.D. 1054 when a new star was recorded, approximately in the position of the Crab Nebula, by Chinese and Japanese observers, the latter estimating it to be as bright as Jupiter; thus there seems little doubt that the object of interest to the oriental observers should be definitely associated with the Crab Nebula.

With these facts in mind Mr F. Hoyle has recently (1945) put forward the hypothesis that the planetary system is a by-product of the explosion of a super-nova. Hoyle assumes that the star which eventually exploded was a member of a binary system, the other component being our Sun; the separation between the two stars is assumed to be about the distance of Jupiter or Saturn from the Sun. These assumptions are not exceptional in any way; binary stars of the kind concerned form, as we have remarked earlier, a considerable proportion of galactic stars, and the appearance of novae is of such frequent occurrence that it is believed in some quarters that the nova-stage is almost a normal experience of the stars in general. Hoyle assumes further that the explosion of the nova is not symmetrical, in other words, that the expulsion of gaseous material is greater in some directions than in others —as it would be if, as seems most probable, the seat of the explosion were anywhere but at the centre of the star; it may be added that, to some extent, the appearance of the Crab Nebula is in accordance with this assumption. We attempt to illustrate the circumstances in Fig. 42.

The components of the original binary star are A and S, the former being the nova and S the Sun. We suppose that the explosion is most intense in the direction AB. Of the vast masses of material shot out from A, a part is eventually captured by the Sun, later to condense into planets. As we have seen in the case of the Crab Nebula the rate of expansion of the nebulous envelope is about 800 miles per second; as it is certain that matter travelling at this speed would not be captured by the Sun, it is supposed that capture occurs only toward the end of the explosive ejection of the stellar material when the speed falls off to a comparatively small value.

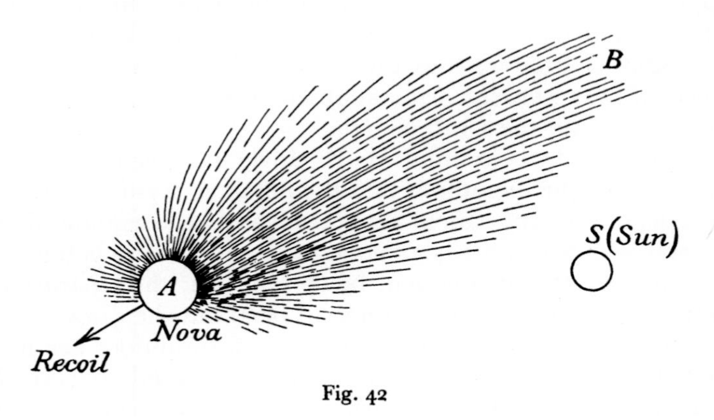

Fig. 42

There is still the problem of the disposal of the nova *A*, since the break-up of the original binary system must be achieved in some way. Just as a gun recoils when a shell is fired, so the explosive ejection of matter from *A*, mostly in the direction *AB* in the figure, will be accompanied by a recoil of the nova *A* in the opposite direction. The velocity of recoil imparted to *A* need not exceed about a score of miles per second to effect the break-up of the binary system. This part of the theory seems reasonable under the conditions envisaged.

Perhaps Hoyle's suggestion as to the origin of the planetary system should be regarded as mainly speculative at present. Although the angular momentum criterion could easily be satisfied, there are still gaps to be filled in the detailed exposition: for example, the explanation of how the rotation was set up in the planets and how the satellites were formed would appear to present difficulties of the first magnitude.

This ends our description of the principal answers to our question 'How?' which have been proposed from the time of Kant and Laplace right up to the present day. In the final chapter we shall attempt to summarize what has been achieved in discussing the subject of the Earth's origin from the various view-points taken up in this and the preceding chapters.

CHAPTER X

EPILOGUE

In this final chapter we first gather together, in condensed form, the results of our discussions on the principal lines of enquiry with which the preceding pages have been concerned, summarizing our attempts to answer the three fundamental questions 'Whence?', 'When?' and 'How?' We also take the opportunity of referring to some very recent investigations which have a bearing on our general theme.

'*Whence?*' In the first section we drew particular attention to the following uniformities observable in the Solar System. First, the planets without exception revolve around the Sun in the *same* direction. Second, the planes in which the orbits of the major planets and of the great majority of the minor planets lie are very much the same. Third, the Sun, the Moon and those planets for which we have observational evidence rotate in the same direction—with Uranus as an exception, although a comparatively small change in the direction of its rotational axis would be sufficient to remove its anomalous character in this respect. Fourth, the great majority of the satellites revolve in the same direction as that in which the planets revolve around the Sun, the exceptions being mainly those satellites at the outermost fringes of the systems concerned. We have seen that there is perhaps a little justification for the suggestion that the satellites not conforming to the majority rule may have been originally minor planets which, through the interplay of planetary and other attractions, have been captured by the parent planets. The reverse process in the case of Pluto—its possible change of status from a satellite to the outermost planet of the Solar System—may be regarded as an argument in support. It may be added that, according to a recent announcement (June 1950), Dr G. P. Kuiper's observations of Pluto with the 200-inch telescope give us more precise information than has been hitherto available for that planet; it now appears that Pluto is a body with a diameter of 3600 miles—almost exactly that of Titan, the largest satellite known—and with a mass estimated to be one-tenth that of the Earth and almost equal to that of Mars.

Fifth, most of the eccentricities of the planetary orbits are small—in such cases the orbits differ very little from circles—and this feature, it is generally agreed, is the result of the 'rounding' of the orbits through the action of a resisting medium in the far-distant past. The eccentricity of a particular orbit is, of course, not constant owing to the gravitational attraction of all the other planets, but the fluctuations are small and circumscribed. However, several minor planets have large orbital eccentricities. One of the largest is that of a minor planet, with an orbital period of $13\frac{1}{2}$ months, discovered in 1949 by Dr Baade at Mt Palomar. At its nearest approach to the Sun this body comes well within Mercury's orbit and at its farthest distance is about 50% more remote than Mars; accordingly, the orbit is extremely elongated, the eccentricity being about $\frac{2}{3}$ according to a rough calculation; also, it is estimated that this body can come within four million miles of the Earth and is probably not more than a mile in diameter. Such large eccentricities suggest that the minor planets concerned have acquired this characteristic through a close approach to one or more of the major planets long after the resisting medium has been swept up.

Sixth, we have seen that about 60 of the terrestrial elements are known to be present in the Sun and that there are very good reasons why the remainder have escaped detection in it. Also, many of the stars are similar to the Sun in chemical constitution. On the other hand we have rather meagre information as to the elements forming the material of the planets and satellites.

The conclusion to be drawn from the uniformities described

is overwhelmingly in favour of a common origin of the Solar System as a whole, many of the anomalies being explicable in terms of special circumstances, such as the interplay of planetary attractions, and, perhaps in several instances, of fortuitous capture. The similarity in chemical constitution of the Earth and the Sun and other stars suggests that possibly the Solar System is the offspring of the Sun, as is assumed in several cosmological theories, or possibly of a star now roaming the galactic regions, as is assumed in other theories.

'*When?*' In the second section we have examined the evidence supplied from a variety of sources. As we have seen, the impasse regarding the age of the Earth in the late Victorian decades between physics and the geological record was finally resolved by the discovery of radioactivity at the end of last century. The age of the oldest known rock is found by radioactivity methods to be not far short of 2000 million years and the most recent work on the salinity of the oceans suggests that the latter are not much behind in age. These results appear to be fortified by the investigations into the ages of meteorites, the oldest of which are estimated to be 3000 million years in age. We shall have occasion later to discuss in a tentative way the influence, on these results, of cosmic rays the investigation of which has been intensified in the past two or three years. Meanwhile, we can summarize provisionally the evidence just alluded to by stating that the age of the Earth is about three, or perhaps four, thousand million years.

The astronomical evidence which we presented in Chapter VIII points to much the same result. The arguments relating to the age of the Sun, the expansion of the Universe and the evolution of the Earth-Moon system appear to lead to conclusions in reasonably good agreement. As regards the last the reader may detect elements of uncertainty in the description of the past history (and the future condition) of our satellite, for much is based, inevitably, on the hypothesis that the rate at which a particular agency operates—such as tidal friction—is dependent on conditions which are assumed to have been much the same throughout the past and will vary little in the future. Again, the argument based on the expansion of the Universe may be regarded perhaps as indicative only of the suggestion that several thousand million years ago, when the Universe was in a highly concentrated state, events such as the formation of the Solar System might plausibly be expected to occur, particularly if an event of this sort is conceived to be of a catastrophic nature. The estimated age of the Sun derived from the consideration of the precise process of evolution now envisaged is satisfactorily in accord with the age of the terrestrial crust and of meteorites or, at any rate, it is of the same order of magnitude.

However we may regard a final result which is perhaps not so clearly cut as some would wish, the achievements of science in the attempt to answer the question 'When?' are indeed very remarkable and, so far as the state of science two or three years ago is concerned, worthy of considerable confidence.

'*How?*' In Chapter IX we examined the principal theories which have been advanced from time to time to account for the birth of the Solar System; in all of these the basic ideas are of a comparatively simple kind and are drawn from a wide range of well-known dynamical and astronomical phenomena. Taking into consideration the problem with which the cosmogonist is confronted it is not surprising that the following-up of an idea—such as the interaction of a wandering star with a binary system, to take only one example—is beset with immense difficulties in mathematical formulation and physical interpretation. Nor is it surprising that no single theory can give an adequate explanation of more than one or two of the many observed characteristics of the Solar System. The scientific search for an answer to our final question has thus failed, up to date, to reach a solution agreeable to astronomers other than those perhaps who, individually, are concerned with the formulation of a particular theory.

As we have seen, many of the principal theories of the formation of the Solar System are of the catastrophic type and the first thing that strikes us about such theories is the 'messiness' associated with them, the contradiction involved in the birth of an orderly planetary system from parent chaotic matter or from the debris of a cataclysmic event. The contrast between the beauty of the heavens and the starkly ugly process, or processes, suggested by cosmogonists constitutes a revolt against our aesthetic instincts, although we must realize that this attitude is beyond the endorsement of whatever logic we bring to bear. On the other hand, a uniformitarian theory such as Laplace's has, despite its fatal weaknesses which are recognized in all quarters, a certain beauty in its orderly development, a feature which was undoubtedly responsible in great measure for its reign as a plausible and long-accredited hypothesis.

In a brief summing-up we can feel some assurance, first, in asserting that the Solar System is derived from a single activity on a great scale in the distant past, though we are unable to decide between the Sun and some unknown star as its material parent; second, in accepting—with some reservations to be referred to later—the conclusion concerned with the age of the Earth in particular; and third, in recognizing that all the theories proposed up to date as to the mechanism by which the Solar System has come into being fail to carry conviction.

In considering the failure just mentioned, it is well not to overlook the limitations of the scientific method. In the present connection the cosmogonist begins by adopting a hypothesis—involving, for example, a rotating nebula or an exploding star—and then working out its implications. The hypothesis must be discarded, or at any rate radically modified, unless it can coordinate all the features of the system which it sets out to explain. The difficulties of substantiating a cosmogonic hypothesis completely are essentially threefold.

In the first place, mathematical analysis—the language in which the phenomena and changes in the material universe are succinctly expressed—is unable at present to cope with the complexity of the problems presented, nor do the fresh developments in astronomy and physics increase the hope that the situation may be alleviated in the future. We are, in fact, still a very long way from the attainment of the perfection implied in Laplace's famous observation, to the effect that mathematicians have finally reduced the entire system of mechanics to general formulae which leave no more to be desired than the perfection of their analysis. The qualification expressed in the last clause and the fact that physics immeasurably transcends the mechanics of Laplace's day in complexity are hardly conducive to a facile optimism that looks for the secrets of the Universe to be soon laid bare.

In the second place the natural laws invoked in support of a hypothesis cannot be considered as immutable as the laws of the Medes and Persians, for they themselves only represent stages in the evolution of scientific thought; the Newtonian law of gravitation and the general theory of relativity represent together perhaps the most notable example of the development of natural law.

In the third place, new discoveries in the realm of physics add complexities to former modes of thought and the emergence of fresh phenomena must have relevance in any comprehensive view of the Universe. As an illustration it is sufficient at this point to mention the recent investigations on cosmic rays with the discovery of mysterious new particles, the mesons, of atomic physics. 'Facts', said Pavlov, the distinguished Russian physiologist, to his students not long before the Second World War, 'are the "air" of the scientist; without them you are unable to soar'; and it might be added that if new facts and new phenomena are ignored the scientist need only expect the fate of Icarus.

We consider for a little the status of natural laws. The laws of physics are of course mainly determined in the laboratory; experiments can be repeated and the conditions controlled or varied at will. We can, for example, study the properties of gases under conditions of constant temperature, thereby eliminating the effects of one variable factor, namely, temperature. In astronomy we are unable to follow the strict procedure of the physicist, for our observations involve many variable factors and one of the main difficulties encountered is that of disentangling the effects of such factors. Further, we cannot go out into space with measuring apparatus to find out if our physical laws are precisely the same in the Andromeda Nebula as on the Earth's surface. If he is to make progress, the astronomer has to assume, in the first instance at least, the universality of natural laws. For example, the constant of gravitation can be determined from terrestrial experiments—we can think of it, without going too closely into detailed specifications, as the force of attraction between two equal particles of given mass at a particular distance apart—and when we determine the mass of a binary system from astronomical observations we use the *same* constant of gravitation. In other words, we immediately introduce the hypothesis that the terrestrial law of gravitation operates exactly in all parts of the Universe and at all times.

Can we be sure, in fact, that the natural law remains unchangeable in space and time? We look out at the Universe and observe its rapid expansion. The distribution of the Galaxies changes from year to year and the average density of matter throughout space is, to all appearances, diminishing. Mach's principle, enunciated near the end of last century, states that the 'inertia' of a body—its capacity to resist force—is conditioned by the distribution of distant bodies, such as the Galaxies, and if this is so it would seem that there must be an evolution of dynamical laws, as indeed follows from the different standpoint of Professor E. A. Milne's theory of kinematical relativity Mach's principle is not accepted in many quarters to-day, but arising partly from it and from other considerations a strange new hypothesis has been introduced to account for the phenomena of the Universe, namely, the 'creation of matter'. Einstein's principle has made us familiar with the doctrine of the equivalence of mass and energy, and in the carbon-nitrogen cycle, for example, we have seen (p. 174) how four protons are effectively transmuted into a helium nucleus, the mass lost in the process finding its expression finally as radiation. Whether or not the reverse process is possible—the transformation of radiation into mass or matter—is not relevant here, for the 'creation of matter' in the present context means exactly what the word 'creation' implies, namely, the appearance of matter out of nothing. This astonishing idea is certainly not shared by Lucretius:* 'Nil igitur fieri de nilo posse fatendum est.'

Very recently Dr H. Bondi and Mr T. Gold, of Trinity College, Cambridge, have put forward a new 'cosmological principle' involving the concept of the creation of matter and based on the following arguments. The observed distribution of the Galaxies, it is believed, is consistent with the homogeneity of the Universe as regards its large-scale features and from this follows the usually accepted cosmological principle that an analysis of the Universe made by observers situated in different parts of space would be independent of their positions *provided* that the observations are made at equivalent times. The new principle, which the authors designate the 'perfect cosmological principle', avoids the necessity for the reservation just made and states that the Universe is homogeneous in its large-scale features and that there is no variability of its physical laws in space *and* time. If we introduce the hypothesis that the laws of physics are to remain valid in an *expanding* Universe—in which the average density of matter is apparently diminishing and space is, so to speak, being

* 'Therefore one must admit that nothing can be made out of nothing.'

'created'—then it is argued that there must be a continuous creation of matter in inter-galactic space so as to restore the balance. A similar hypothesis has also been followed up by Mr F. Hoyle from a different standpoint. According to Bondi and Gold the rate at which matter is created—or spontaneously appears—is extremely minute; a calculation, depending on the observed rate of expansion of the Universe and on the average density of matter within it, as now estimated, shows that the time required for matter, equal to the mass of the Earth, to be created within a sphere of radius equal to the distance between the Sun and the nearest star (that is, 25 million million miles) would be about 70 million years.

I have very briefly described this new theory, partly for its intrinsic interest, partly to illustrate the changing flux of ideas with regard to the spatial, temporal and material characteristics of the Universe, and partly to show the dependence of a theory on a basic hypothesis which, incidentally, may or may not be true. The consequences of the theory must eventually be tested against observations and, even if the accordance appears to be satisfactory, there is still no certitude that some other theory, based on a different hypothesis, will not answer just as well.

Reference has already been made to the possible necessity of modifying theories when new phenomena are discovered or when certain phenomena cannot be satisfactorily explained by existing theories. For example, the Newtonian universe, bound by the familiar law of gravitation, has had to yield to the relativistic universe with its large-scale expansion inferred from the observed recessional velocities of the extra-galactic nebulae. In the early post-war years an immense amount of research has been undertaken in the investigation of cosmic rays and of the properties of mesons to which these rays give rise. The first hint of the existence of an unknown extra-terrestrial radiation was given as far back as 1900 by Professor C. T. R. Wilson who found that, when all precautions were taken, a charged electroscope gradually lost its charge and this was shown to be due to the ionization of the air within the sealed enclosure. Wilson at first suggested that the ionization was produced by some unspecified radiation arriving from outside the atmosphere; but this idea was later allowed to lapse when observations, made in a railway tunnel, gave no perceptible diminution of the effect, contrary to what was expected from the shielding of the instrument by a thick layer of rock and earth. In later years, however, it became apparent from observations carried out at heights as great as five miles above the Earth's surface and at great depths below the surface that the so-called cosmic rays are indeed a reality. As there was no sign, at a given locality, of any variation of the effects at different hours of the day and night, it was concluded that the Sun and the Milky Way must be dismissed as possible seats of the origin of the rays; it was then assumed that they must come from the 'cosmos' itself—hence the name.

In 1938 new particles called *mesons* were discovered by Dr Anderson at Pasadena. The nature of these particles was not clear until 1947 when Professor Powell of Bristol showed that there were two types of meson of comparable mass—π-mesons of about 276 electron-masses and μ-mesons of about 212 electron-masses. Only the π-mesons interact strongly with nuclei and are believed to be largely responsible for nuclear forces. The μ-mesons arise mainly as decay-products of π-mesons. Recently, the existence of several varieties of heavier mesons has been established. Amongst these are the V-particles first observed by Rochester and Butler at Manchester in 1947. It may be added that mesons were predicted, on theoretical grounds, by Yukawa in 1935.

Mesons have great penetrating power and it is these that are responsible for the phenomena observed at or near the Earth's surface. The cosmic rays which come from extraterrestrial regions and which impinge on the Earth's atmosphere are believed to be high-speed protons and electrons which, interacting with the atomic nuclei in the upper levels of the atmosphere, produce the penetrating mesons. How the cosmic rays originate is not known, nor has any coherent theory of mesons yet been formulated; evidently, however, we are on the threshold of a new and, perhaps, revolutionary interpretation of matter.

The fact that all space now appears to be filled with cosmic rays must make us cautious in pressing some of our earlier conclusions. It is not very easy, of course, to see how cosmic rays can affect, if at all, any proposed hypothesis as to the formation of the Solar System or the evolution of the Earth-Moon system. But may it not be relevant to ask what impact the new discoveries have on our assessment of the age of the Earth and of meteorites in particular? We have seen how elements are transmuted into different elements through bombardment by protons and neutrons. In deducing the ages of rocks and meteorites from measurements of radioactive decay we may be omitting some very relevant factor associated with cosmic rays—just as Kelvin's arguments took no cognizance of the then-undiscovered phenomenon of radioactivity. For example, meteorites roaming about interplanetary space must be sustaining a continuous bombardment by cosmic rays and it is more than likely that the production of helium occurs on a much more rapid scale than is contemplated through spontaneous radioactive transformations; consequently, it would seem that the true ages of meteorites are not accurately assessed by the uranium-helium method. Also, it may be possible that what we call the 'spontaneous disintegration' of radioactive elements is related in some way to the action of cosmic rays and, if so, the rate of disintegration may vary from century to century according to the intensity of the rays. This is merely one example in which new facts and new knowledge are likely to influence what had been regarded a year or two ago as firmly established conclusions.

The origin of the elements is another problem to which attention is being directed at the present time. In dealing with the methods of liberation of energy within the Sun and stars we have seen that a supply of one or more elements heavier than helium is requisite for the processes envisaged. It would appear that helium is the only element that is fabricated out of the plentiful supplies of hydrogen within the Sun and stars. Where have the heavier atoms been formed such as lithium, carbon and nitrogen, which appear in the transformations described earlier, or have they existed from all time past? It is perhaps pertinent to remark here that several elements have been artificially produced in the laboratory which are heavier than uranium, the heaviest 'natural' element found on the Earth. Have such trans-uranium elements, as they are called, existed in the past, perhaps formed out of simpler elements under conditions vastly different from those we recognize in the Universe at present? Several attempts have already been made to answer such questions; some are based on physical conditions involving thousands of millions of degrees of temperature, and others on conditions of density amounting to a million million times that of water, together with not inordinately high temperatures. Such conditions, if they ever existed, must have occurred long before the stars and the Galaxies were formed. Perhaps an answer may be found some day, but at present those who investigate such recondite matters are cognizant of the high degree of speculation which their work involves.

It has been the fashion in recent years for men of science to enter the field of philosophy—the reader need not feel any apprehension here—sometimes with profit to both science and philosophy. This is as it should be; for the layman tends to be impressed only by the achievements of science and is frequently ready to accept uncritically any new theory or speculation, perhaps not understanding very much about it. Some of the reasons for this attitude are related to the prestige of the scientist, founded on the spectacular and authenticated discoveries particularly in astronomy, to the precision of measurement now attainable in many fields, to the endorsement of mathematics, and so on. The layman, it would seem, is disposed to agree with Sir J. M. Barrie who was credited with the observation that 'the man of science appears to be the only man who has anything to say just now', although he might not quite agree with the rather unkind addendum 'and is the only man who does not know how to say it'. In defence, it might be argued that it is only the man of science who fully realizes the immense difficulty in trying to explain the new and bizarre developments of recent years and, if one is allowed to judge by the widespread and continuous interest in all matters scientific, it would be unfair to say that his expositions have been completely ineffective. Perhaps this interest is mainly a reaction from the apparently insoluble problems of the post-war years, just as in 1919 the strange new theory of relativity stirred the imagination of a war-weary world in which the ramparts of ethical and spiritual values appeared to have crashed like the walls of Jericho. The authority of the scientist in his own province is unquestioned, for his activities are governed by his own well-established rules of measurement, by the formulation of theories and by the search for order in Nature; the danger lies in believing that this authority can be transferred to other realms of thought. In a cricket match we almost invariably accept the authority and the decisions of the umpire in applying the appropriate rules; but we would firmly contest the suggestion, except in exceptional cases, that he would necessarily be an equally safe guide in discussing the rules of life, the problems of political philosophy, of ethics and of religion.

Fortunately, scientists are becoming more aware of the dangers inherent in a blind belief that science is a golden key which can open the door to *all* knowledge and are ready to agree that there are far greater, or more important, problems in the field of human experience than those to which the scientific method is appropriate.

Our study in this book has been concerned with inanimate matter ranging from the nimble electron to Galaxies of unimagined grandeur. The achievements recorded in the previous chapters have been many and even, on occasions, spectacular and certainly beyond disparagement; nowhere, however, have we touched the greatest topic of all—man's place and destiny in this marvellous creation. Our study has been one of the stage only, and we have learned much about its construction, properties, lighting and so on, but we have been in no position to investigate the characters, the aspirations or even the foibles of the actors and, most important of all, the mind of the author, in the background, who has created the play.

When we study the Universe and appreciate its grandeur and orderliness, it seems to me that we are led to the recognition of a Creative Power and Cosmic Purpose that transcends all that our limited minds can comprehend. In one of his essays Francis Bacon expressed this belief picturesquely as follows: 'I had rather beleave all the Fables in the Legend and the Talmud and the Alcoran than that this Universall Frame is without a Minde'. To-day we have learned very much more about the 'Universall Frame' than was known in Bacon's time; nevertheless, to many of us, scientific and non-scientific alike, the belief in a Divine Creator is as necessary now as ever it was. To one astronomer at least 'The Heavens are telling the Glory of God and the Wonder of His Works'.

INDEX

Printed in the United States
By Bookmasters